The Woman in the Body

The Woman

Emily Martin

in the Body

A Cultural Analysis of Reproduction

Beacon Press / Boston

Beacon Press
25 Beacon Street
Boston, Massachusetts 02108

Beacon Press books
are published under the auspices of
the Unitarian Universalist Association of Congregations.

First published as a Beacon Paperback in 1989

Printed in the United States of America

94 93 92 91 90 89   8 7 6 5 4 3 2

Library of Congress Cataloging-in-Publication Data

Martin, Emily.
The woman in the body.

1. Women—Physiology. 2. Women—Psychology.
3. Women—Health and hygiene—Sociological aspects.
I. Title.
RG103.5.M37 1987 306'.46 86-47754
ISBN 0-8070-4604-3
ISBN 0-8070-4605-1 (pbk.)

Text design by Chris L. Smith

For Jenny and Ariel

Contents

Acknowledgments

This study has rested on the backs of many people besides my own. Foremost are the women who consented to be interviewed, giving their time, spilling their tears, sharing their insights, and opening their minds and hearts. Their collective efforts are the heart of this book and give it its best chance of growing in the hearts of other women. No less important are the women who conducted many of the interviews: Sarah Begus, Betsy Getaz, Ewa Hauser, Jane Sewell, Susanne Siskel, and Andrea Taylor. For enduring the difficulty of approaching strangers, the agony of trying to find addresses in unfamiliar neighborhoods, the fear of standing in front of a group and asking for help, and for believing in the possibility that ordinary women will have wisdom, I thank them. For offering me their insights about what the women said in the interviews and for offering the women they interviewed open, receptive, unjudging ears, I cannot thank them enough.

The women we interviewed must remain anonymous, but those who helped us find them, allowing us to explain the project to groups of women within their center or organization, need not. I am particularly grateful to Tom Culotta, Beverly Daniels, Susan Doering, Marcia Hettinger, Mary Jane Lyman, Eleanor Mann, Cindy Monshower, Elizabeth Reiley, Phillip Schmandt, Jane Szczepaniak, and Anita Wormley. The most laborious work was done by several undergraduates who transcribed tapes, searched for references, and performed many editorial chores. For this, I thank Soccoro Alcaden, Sharon Crockett, Lora Holmberg, Jonathan Lewis, Carol Lundquist, and Leslie Nuñez. The resourceful, efficient librarians in the Interlibrary Loan Department at Milton Eisenhower Library, together with the swift couriers of the Eisenhower Express, put hundreds of books and articles on my desk, from obscure or rare medical texts to the latest popular tracts. I also appreciate help given by the librarians of Enoch Pratt and Welch Medical Libraries and by Doris Thibideau, Rare Book Librarian of the Institute of the History of Medicine. Generous financial support, which made it possible for me to enjoy assistance in the research as well as a year of writing, came from an American Council of Learned Societies Fellowship and BRSG Grant S07 RR07041 awarded by the Biomedical Research Support Grant Program, Division of Research Resources, National Institutes of Health. Portions of Chapter 5 appeared as an article in *Social Science and Medicine,* 1984, 19(11):1201–6.

Over the past several years, many people I have known in Baltimore in study groups on feminism, Marxism, and science have introduced me to literature I would otherwise not know and critical perspectives from disciplines other than my own. Most important among them have been Sarah Begus, Toby Ditz, Elizabeth Fee, Sandra Harding, Nancy Hartsock, David Harvey, Sharon Kingsland, Karen Mashke, Vicente Navarro, Kathy Peiss, Jo-ann Pilardi, Kathy Ryan, Jane Sewell, Marylynn Salmon, Dan Todes, and Karen Olson.

I owe a special debt to my colleagues in the Department of Anthropology at Johns Hopkins University—Gillian Feeley-Harnik, Ashraf Ghani, Sidney Mintz, and Katherine Verdery—for their warm enthusiasm about this project and their help relating my findings to anthropological issues. Several people read earlier versions of portions of the manuscript, and I thank them for their encouragement and criticism: Carol Browner, Elizabeth Fee, Alma Gottlieb, Sandra Harding, Ruth Hubbard, Judith Leavitt, Leith Mullings, Lucile Newman, Rayna Rapp, Barbara Katz Rothman, Carroll Smith-Rosenberg, Michael Taussig, and Linda Whiteford. Thanks also go to Joanne Wyckoff at Beacon Press for her encouragement from the early stages of this project and for her sound editorial advice, to Barbara Flanagan for her excellent copy editing, and to my eleven-year-old daughter, Jenny, for thinking up the title.

To those who not only gave help with the contents of the book but fortified my spirit with emotional support, particularly during the last year of writing, I am deeply grateful: Sarah Begus, Jane Eliot Sewell, Lorna AmaraSingham Rhodes, and Richard Cone. Without the support of my husband, Richard, this book truly could not have been written. Out of his generosity of spirit he gave me not only unflagging courage and tender appreciation but—a gift few are capable of—willingness to subject his own scientific expertise to analysis not only as science but as culture. Through their loving and reliable child care, Virginia Miller and Jane and Mike Sewell provided my peace of mind. My daughters gave me warm arms to come home to and hope for the future. In hope, I dedicate this book to them.

I know no woman—virgin, mother, lesbian, married, celibate—whether she earns her keep as a housewife, a cocktail waitress, or a scanner of brain waves—for whom her body is not a fundamental problem: its clouded meaning, its fertility, its desire, its so-called frigidity, its bloody speech, its silences, its changes and mutilations, its rapes and ripenings. There is for the first time today a possibility of converting our physicality into both knowledge and power.

—Adrienne Rich
Of Woman Born

Problems and Methods

1 The Familiar and the Exotic

When automobiles replace outrigger canoes, apples replace mangosteens and papayas, and boxes of chocolate are substituted for octopus pudding as a suitable gift to take on a date, the whole incidental paraphernalia of strangeness which I could use to convey the cultures of other peoples is gone.
—Margaret Mead
And Keep Your Powder Dry

Until recently, many anthropologists looked with suspicion on the task of trying to understand one's own society. The traditional and proper norm for our field was to study people in a foreign, exotic society, living as a participant-observer in their midst, learning a strange language and adopting unfamiliar ways as much as possible, until, by both constant immersion and some degree of empathy, one could make comprehensible to oneself and one's professional colleagues what were at first incomprehensible customs and behavior. This was the kind of fieldwork I did as a graduate student and young assistant professor in Chinese villages in Taiwan.

I think that the seeds of my desire to try doing fieldwork in my own society were sown in that foreign place, as a result of questions Chinese villagers frequently asked me about the United States. I could give superficial answers to questions about why we put our aging parents in old people's homes (a custom that made us seem little better than animals in the eyes of the Chinese), why I still had no children at the advanced age of 25, why the child I brought with me at a somewhat later age might well be my only one, even though she was a girl—but I realized that at a deeper level I really did not have good answers.

In the years after my fieldwork in Taiwan, the problems anthropologists regarded as central began to undergo change. Far from our earlier suspicion that fieldwork at home was unacceptable because it did not make the exotic become familiar, many anthropologists began to feel that we urgently needed cultural studies of western societies. As anthropologists began to study the impact of western factory production, wage labor, and commodities on far-flung societies, we realized we needed to understand better the nature of what we were exporting.[1] At a deeper level, we began to see that for all our intention of understanding other cultural worlds, as long as our own assumptions—about gender, say—remained unanalyzed, they could creep unnoticed into accounts of other places.[2] We were in danger of treating foreign customs and beliefs as in need of explanation or translation and our own as self-evident or, for those aspects of our beliefs we label "scientific," true.

When it came right down to starting a study in the United States, not much of what I had learned in Taiwan helped. I knew I wanted to study how general cultural processes affect women, and I wanted to encompass ethnic and class differences. But how was I to find a residential community or occupational category that captured those differences? Clifford Geertz has urged us to apply anthropological techniques to the study of modern thought by looking for natural communities in our midst, people who share involvement in one another's lives, even if the involvement is the intellectual sort common to mathematicians or theoretical physicists.[3] But there is no community on earth, even a thought community like Geertz's "intellectual villagers," that could represent the experiences of women, living as they do in all class strata and nearly all community and occupational groups.

Even though women in our society do not form a face-to-face or word-to-word community, they do share some experiences: all are defined as "women," one of two usually permanent gender categories to which everyone in our society must be assigned; all (some more than others, some more aware than others) occupy subordinant positions to men, if not in their jobs, then in their families, and if not in their families, then in general cultural imagery and language; all have female bodies and experience common bodily processes such as menstruation and childbirth (however various the meanings such diverse groups as black Americans or Jewish Americans can give these processes); all are affected in one way or another by medical and scientific views of female bodily processes.

In the end, I decided to conduct extended interviews with women in many different communities, hoping to see whether women's common experiences take different forms and meanings in different social and economic conditions. Since my overarching concern was the social production of knowledge, I had to be sure the social circumstances of the women I interviewed varied significantly. To achieve this end and to capture the experiences of women in different phases of the life cycle, my original research design was to interview 165 women, about 55 in each of three life stages—after puberty but before childbearing, during the childbearing and childrearing years, and during and after menopause—making sure that about half of each group was clearly middle class or above and the other half working class or below. This plan would not allow me to make fine distinctions among the multitude of different class positions in society, but it would allow me to see whether, at the level of crude differences, class position mattered.

Baltimore's social organization, which includes many traditional neighborhoods, helped me incorporate class differences. By choosing community centers or health clinics in Remington, a neighborhood near the city's center, I could reach mostly white working-class women; going to Roland Park, toward the northern city limits, was practically a guarantee of the middle or upper middle class; and by visiting Park Heights in the northwest section of the city I would talk with black working-class women. The edges of the clusters "middle class" and "working class" are fuzzy, as they should be for groups formed on the basis of complexly overlapping characteristics in a state of constant change. For the purposes of this study I take "middle class" to include: self-employed businesspeople, white-collar or blue-collar salaried professionals and managers, salaried administrative sales and clerical employees, supervisory employees, fire fighters, and police. For the most part these occupations give workers greater autonomy, responsibility, security, mobility, and prestige than working-class occupations.[4]

Although a broad grouping like this—from almost upper class at the top to almost working class at the bottom—must contain a great variety of family and community forms, I have found Rayna Rapp's attempt to specify general traits common to many middle-class families very useful.[5] According to Rapp, since middle-class households can count on a relatively secure resource base, they can spend some resources on recreation or luxuries. When emergencies occur, they call on institutions outside the family: banks, pensions, or credit accounts. Their greater geographic mobility (by choice or as a requirement of keeping a job) means that families must rely on purchased services rather than kinship or community help on a day-to-day basis and especially when in transition. Marriages are expected to involve communication and companionship between spouses, and women are expected to choose whether to fill only the roles of housewife and mother or to also find a job outside the home for self-fulfillment or additional income. Because resources are largely accumulated to be passed on to the next generation, families emphasize lineal kinship links from parents to children to grandchildren. People look to friends for affective ties outside the family, but they do not usually share significant resources with friends, to avoid diverting funds laterally instead of lineally.

"Working class" I take to include skilled craft workers, clerical office workers, retail sales clerks, factory laborers, and service workers.[6] Although some of these occupations are white collar and more prestigious than others, all have suffered increasingly the same loss of workers' autonomy and control as factory jobs have.[7]

In contrast to middle-class households, working-class households, relying on a more tenuous resource base, tend to develop many forms of pooling and sharing within neighborhoods and extended families. Women are ideally seen as guardians of the home, which is regarded as an autonomous place different from the workplace.[8] Within this domain, women are charged with the task of purchasing goods so that wages are stretched to meet needs; in this task they often rely on sharing and exchanging laterally, among neighbors and kin. Of course, the reality of dependence on wage labor often sends women out of the home for work, primarily into the sex-segregated lowest-paid sectors of the economy.[9]

I feel confident that the total set of interviews captures significant class differences, even though it certainly does not reach to either the very top or the very bottom of the class hierarchy. Most of the direct questions we asked were related to health matters, and what the interviews show is class refracted through concerns about health. Since in our society one's household's resources (such things as income, savings, education, or experience dealing with doctors) dramatically affect one's diet, one's opportunities for exercise, and one's access to health care, drastic differences on these matters are evident from the very first exchanges. The first questions in the interview usually included "Who lives in your household or family? What do they do? What is done in your family for the sake of health?" Here are some examples of contrasting responses:

I am a senior in college majoring in natural science. My mother is a psychotherapist, my father is a computer systems analyst. My sister is a sophomore in college. My parents exercise almost every day because heart disease runs in the family, and my father has heart disease. We have a bicycle and a rowing machine, so he does aerobic exercise pretty much every morning but the rest of us are basically lazy and overweight. [If you get sick?] You go to bed early, take lots of aspirin or Tylenol, drink lots of fluids, stay warm. We were never ones to run to the doctor at the slightest sign of fever or cough. It was usually just stay in bed. (Rachel Lehman)[10]

I'm a junior in college, not big on exercise. We used to go skiing, but it wasn't done for the sake of exercise. Mother writes a food column; she's always cooking and eating unusual things. If we get sick we're taken to the town family doctor and urged to drink orange juice, sleep—your basic all-American prescription. Dad is a self-employed architect, remarried, right now on a pilgrimage back to Israel. (Mara Lenhart)

As I was growing up, my family felt very proud about being able to eat meat. It was a big deal to be able to afford meat. When I was very young, I had a sense of not being clean, of not being taken care of. And I wasn't. I think some of that was my mother's background. Some of it was the reality of our life. It was so hard. She had four children. She was divorced and she was trying to support us all by herself. She had no education. She went to the eighth grade. She was a barmaid for a while, and she didn't make much money. It was real tough. When you have four kids, it's hard to keep after them. It's hard enough to keep after one. When I was young, I remember making up what I had for breakfast at school because I didn't have any breakfast. (Ann Morrison)

I'm in tenth grade. I want to be a doctor; I want to be a specialist dealing with the heart. My mother's a liaison worker, my father's a missionary attendant. [What's a liaison worker?] She works in a Title I school and has to make home visits to student houses when they don't come to school. People at the school get together food baskets; when somebody's house burns down, they collect clothes and give them to them. I have two sisters and one brother. My father, my brother, and my little nephew live with us—my nephew is my sister's son. My sister lives down the street from us. Because she works as a beautician, her son lives with us. My mother and my brother are supposed to go on a diet, me and my father, we're not on one. Sometimes I exercise, but my legs get sore because I ain't used to it. When I was at [another] high school as a cheerleader, I was doing exercise then. [When a family member got sick?] Go to the hospital! I just go to my mother and she'll take care of me, if it's real bad she'll take me to the hospital. (Carmelita Baker)

I present these examples out of context, but in broad strokes they reveal wide contrasts in how health relates to household resources: family doctors, special exercise equipment, certain expensive foods on the one hand, and hospital emergency rooms, haphazard exercise, ordinary foods on the other. This, of course, only scratches the surface because work conditions, housing conditions and the hazards associated with them, access to health insurance or saving for medical emergencies vary radically among these women's families.

The women represented in this book are self-selected rather than randomly sampled. We found women who were willing to participate by explaining the project to them in small groups (exercise classes, school classrooms, childbirth education classes, senior citizen programs, churches, community organizations, health clinics) and asking for volunteers. In this way and by pursuing other women suggested by the volunteers, we built up the numbers. We conducted 165 interviews, 29% of them in the youngest life stage (puberty to childbearing), 42% in the middle life stage (childbearing and childrearing age), and 29% in the oldest life stage (menopause and postmenopause). Overall, 43% were working class and 57% middle class. Of all these, 28% were members of Baltimore's (and the nation's) largest ethnic minority, black Americans. (See Appendix 2 for more details on all of the women interviewed.)

We conducted almost all the interviews in women's homes, although occasionally for convenience we did them in a room provided by a school or other organization. The actual questions we used as a guide are in Appendix 1. We tape-recorded with permission in all cases and transcribed tapes into an information processing system that indexed every word. This freed me from the constraints of memory and the labor so familiar to fieldworkers of preparing a (comparatively schematic) index by hand. At a moment's notice, I could request my computer to pull up every paragraph in which any particular word or combination of words had been used.

The interviewing itself was very much a team effort. The several research assistants who helped do the interviews were all candidates for advanced degrees in the social or health sciences. By listening to each other's interviews and discussing our individual problems with specific questions and the anxieties we suffered asking them, we managed to support each other and keep the interviews consistent. By and large we tried to match interviewer and interviewee: most of the youngest white women were interviewed by a white interviewer in her early twenties; most of the youngest black women and many of the black women in the middle age group were interviewed by a black interviewer in her twenties who has a new baby at home; most of the white women in the middle age group were interviewed by me, 37 years old and pregnant with my second child when the study began; and most of the older women were interviewed by an interviewer in her forties, whose children are teenagers. These were not hard and fast rules: as the one who had to make sense of all this, I did some interviews in every category, and I found that only the white middle-class teenagers seemed to feel uncomfortable, perhaps because I could well have been a mother to them.

Our best luck finding people who wanted to participate was through small organizations or large ones with local autonomy, especially where we could be introduced through someone we knew personally. Large organizations often required cumbersome clearances designed to provide access to far more people than we could ever reach. Complex reciprocal relationships developed for many of us, in which we gave back what we could to individuals, in the form of information, referrals, help dealing with government agencies, or simply a sympathetic ear, and to local organizations in the form of public talks or help writing grants. Ultimately, of course, as in any fieldwork, the gifts of insight that people gave us into their lives can never be repaid. But many people were willing to give that gift in the hope that through greater understanding of the problems women face, things might change.

Doing a study based on interviews meant that I gave up the rich, multilayered texture of life that I would have experienced by living in a community or with a family. I tried to make up for that by participating in as many ongoing organizations as possible, such as organizations for childbirth education or for the prevention of cesarean section. As it turned out, doing fieldwork through interviews was far less abstract than I feared. Although I initially felt I was doing fieldwork only episodically, rather than almost constantly as in Taiwan, the episodes could be very intense. All of us doing interviews often felt swept away by them—either exhilarated or cast down—and the emotional effects lingered, as if we had had the most profound events of someone else's life shoehorned into our own.

And these other lives did not stay compartmentalized. Friendships developed—not always, but often—and when women I had interviewed invited me to an Easter egg hunt, a Christmas party, or to a rock band's performance of a song about premenstrual syndrome (PMS), it felt a lot like fieldwork of the traditional kind. As the study went on, I found it easier to integrate the various parts of my life, sending women books about issues that concerned them or going to an academic women's studies conference at a nearby university with a local birth activist.

Nor would the issues stay out of my life. Whereas in Taiwan I was able to periodically escape out of the foreign setting into a familiar one (reading *Time* magazine or watching an American movie would do this), in this project there was nowhere to escape. The things I was discovering about gender in our society and about the lives of other women overlapped and informed both the most personal aspects of my life—how to arrange care of my children while I worked, how to live in an egalitarian way with my husband—and the most public—how to achieve a semblance of gender equity in my university.

When I began to try to make sense out of the interviews, one of the deepest differences between doing fieldwork in my own society and elsewhere became apparent. Over and over again, what women told me seemed at first like so much common sense. For example, I began my interviews with women who were pregnant or who had recently given birth. As I discuss in Chapter 4, women talk as if uterine contractions were separate from the self and as if labor were something one went through rather than actively played out. But since all the medical texts and the prepared-childbirth literature I was reading were telling me that the uterus is an involuntary muscle, I had no sense of puzzlement about what either the texts or the women were saying. In the light of the medical texts' description, the locutions and images of my informants seemed perfectly fitted to reality. If uterine contractions are involuntary, then of course women talk about them as if they are distinct and separate from the self, and of course women describe labor as something that one goes through rather than something that one does. For many weeks I felt only a sort of leaden disappointment that all those interviews had turned up views of the body that reflected no more than actual, scientific fact.

The realization that statements about uterine contractions being involuntary are not brute, final, unquestionable facts but rather cultural organizations of experience came to me as a sudden and complete change of perspective. All at once I saw these "facts" as themselves standing in need of explanation, the way I nearly always saw statements of "fact" about the body made by Chinese villagers in the field: this is a "hot" illness, your yin and yang are out of balance, and so on. The length of time it took me to make this shift stands as vivid testimony to how solidly entrenched our own cultural presuppositions are and how difficult it is to dig them up for inspection. The one I stumbled over was my acceptance of scientific, medical statements as truth, despite many warnings I had made to myself and heard from others about precisely this kind of danger when one tries to do fieldwork in one's own society.

Even more striking, I anguished over the obviousness of everything the women were saying. Marx explains how people do not notice contradictions in their own society: "A complete contradiction offers not the least mystery to them. They feel as much at home as a fish in water among manifestations which are separated from their internal connections and absurd when isolated by themselves."[11] That women's responses in our interviews were obvious to me is a way of saying that I felt as much at home hearing them as a fish is in water. As an anthropologist, my problem was how to find a vantage point from which to see the water I had lived in all my life. Berger and Luckmann have expressed this problem as "trying to push a bus in which you are riding."

Not everything about fieldwork in the United States was more difficult than fieldwork abroad. Many women were extremely forthcoming and easy to interview, perhaps partly because the cultural form of the recorded interview is familiar and acceptable from TV and radio talk shows. Similarly, the customs of doing research, publishing books, and even attaining a critical stance on established institutions could be taken for granted as acceptable cultural practices. In addition, especially with middle-class women in Baltimore, my affiliation with Johns Hopkins University was usually an advantage. Johns Hopkins is a prestigious school in the area, and pride and identification with its doings, from medical inventions to lacrosse victories, is common.

Not always, though. One black woman declined to be interviewed because she did not like studies that Johns Hopkins' medical school had done in the poor populations surrounding the university in the inner city, and one administrator in a predominantly black high school located near Johns Hopkins Hospital refused me access to the students, saying, "We do not want to get involved in any study. This high school has been interviewed and interviewed and interviewed."

In the interviews we did, women talk mostly about menstruation, birth, menopause, and so on. In this book, my goal is to convey a sense of underlying cultural assumptions about these events, a sense of their implicit meaning. To anticipate a likely misunderstanding, since I am talking about events we conceive of as biological, and often medical, it may be tempting to say: it is all very well to elaborate on premenstrual syndrome (or labor or menopause) as a cultural form whose meanings are constructed out of people's different life situations, but the fundamental basis for events like these is biological, hormonal, genetic, or whatever. In their book about the new biological determinism, Lewontin, Rose, and Kamin argue that rather than either biological or cultural determinism, we need an "integrated understanding of the relationship between the biological and the social."[12] Rather than breaking wholes down into parts and deriving the properties of wholes from those parts, often including the attempt to assign relative weight to various partial causes (PMS is 23 percent hormonal, 45 percent psychological, and so on), what we need are accounts in which the "properties of parts and wholes codetermine each other."[13]

To explain this they use the example of baking a cake:

> Think, for example, of the baking of a cake: the taste of the product is the result of a complex interaction of components—such as butter, sugar, and flour—exposed for various periods to elevated temperatures; it is not dissociable into such-or-such a percent of flour, such-or-such of butter, etc., although each and every component (and their development over time at a raised temperature) has its contribution to make to the final product.[14]

The example would be much more enlightening if they had added that the cake was chocolate, baked by a 64-year-old widow on the occasion of her only daughter's departure for an extended residence abroad. Adding these complexities, we would be hard pressed to say how much the bittersweet taste of the finished cake was due to the chocolate and how much was due to the social significance of the occasion.

I assume that women would not menstruate or give birth if they did not have physical bodies and if those bodies did not contain genes, hormones, and many other things. Questions about how genes or hormones function in human lives are legitimate, but even complete understanding of them could never settle any human matter. If we focused only on completely understanding the physical components of the cake, we would lose the person who baked it, the occasion it was baked for, and the people who were there to eat it. In this book, even though I talk at length about biological and medical processes, my concern is not what is true or false about those processes, nor am I competent to say. Instead I try to get at what *else* ordinary people or medical specialists are talking about when they describe hormones, the uterus, or menstrual flow. What cultural assumptions are they making about the nature of women, of men, of the purpose of existence? Often these assumptions are deeply buried, not hidden exactly, but so much a part of our usual experience of the world that they are nearly impossible for a member of the same cultural universe to ferret out.

Trying to do this kind of study while being a professor at Johns Hopkins, I often felt like a mouse in the den of a lion, and a disrespectful mouse at that. Hopkins is of course a citadel of excellence in medical education and practice, as defined by the medical profession. Historically Hopkins' medical school has been chosen as the model on which all other accredited medical colleges have been based. But this was part of a process of tightening the professional organization of medicine, both limiting its numbers to reduce competition and restricting them largely to white males.[15]

Although I will be critical of many central ideas current in medicine because I think they are demeaning to women, doctors as individuals are certainly not to be held responsible. Medical culture has a powerful system of socialization which exacts conformity as the price of participation.[16] It is also a cultural system whose ideas and practices pervade popular culture and in which, therefore, we all participate to some degree. There are no individual villains in this book. Although doctors, like anyone else, can make mistakes, and I was sometimes told about them in interviews, I have steadfastly not used any of those cases to illustrate my points: I want to get at normal, correct, routine medical procedures and the ideas behind them.

I have explored current medical ideas by focusing on texts that are the basis of teaching in medical schools and handbooks that are guides to practice in hospitals. In addition, I have listened to lectures for premedical and medical students and to many casual conversations with colleagues who are doctors or biological scientists. However, my data are unbalanced in one way: I do not have the same kind of rich interview material for doctors that I have for ordinary women. This material would be very valuable, and in fact is in the process of being gathered by other researchers. Margaret Lock has shown how clinical settings of different kinds—especially progressive, family-centered ones—can ameliorate the hard and fast views of medical texts.[17] Although I am sure she is right about variation in medical practice, I have tried to keep to the level of the "grammar" that scientific medicine uses to describe female bodies, and I am confident that the deep level at which such a grammar is formulated and transmitted means its terms are not easily forgotten or dropped. We will see that the consequences of the medical lexicon about women's bodies show up clearly in statistics, such as the rate of cesarean section, and vividly both in women's perceptions about how medicine views their bodies and in how women view their own bodies.

2 Fragmentation and Gender

One task of human thought is to try to perceive what the range of possibilities may be in a future that always carries on its back the burden of the present and the past.
—Barrington Moore
Reflections on the Causes of Human Misery

What are the main categories by which we in the United States think and act in the world? How do occupants of particular places in that world—such as women in different social and economic positions—see these categories? Can we speak of one homogeneous view of the world, or do things appear very different if one looks with the eyes of a woman? A middle-class woman? A working-class woman? A black woman?

Let us begin by looking at how our major social categories have been described by social historians and anthropologists at the level of the social whole, at the level of the "person," and at the level of the "body." We will see that a dominant theme in studies of modern representations of the world is that these divisions constitute fragmentations and pieces of something that once was whole or would be better if it were whole.

Social historians looking at the family in western societies have identified some sharp separations that emerge with particular intensity in the development of capitalist societies, most centrally the separation of the world of work from the world of family life, of the public sphere outside the home from the private sphere inside the home.[1] Our lives have come to be organized around two realms: a private realm where women are most in evidence, where "natural" functions like sex and the bodily functions related to procreation take place, where the affective content of relationships is primary, and a public realm where men are most in evidence, where "culture" (books, schools, art, music, science) is produced, where money is made, work is done, and where one's efficiency at producing goods or services takes precedence over one's feelings about fellow workers.[2]

These divisions arose during industrialization, as "the size of productive units grew and productive activity moved out of the household to workshops and factories. Increasingly, people worked for wages."[3] But whereas working-class women followed wages into the labor market, middle-class women retreated into the home:

> A cult of domesticity demanded that the bourgeois female cultivate the gentle arts of femininity. The leading characteristics of femininity were abstinence—both abstinence from labor and abstinence from sexuality—and reproductivity, that is, the production of children. "The functions of the wife," went one formulation, "except among the poorest class, are or ought to be exclusively domestic." That meant she should "bear children, regulate the affairs of the household, and be an aid and companion to her husband." Her social importance lay in her very idleness. Nonproductivity was a major indicator of class standing, a working wife a sign of social and economic disaster.[4]

The two realms are not regarded equally: success in the public realm is nearly the only road to high social positions, most of which are held by men; being productive in the world of paid work (except for menial jobs) counts for more in the dominant cultural view than whatever one does as a "dependent" in the domestic sphere. Because the realm of work historically involved a break from nature and entailed an effort to dominate nature, women, those associated with the "natural" realm of family, were seen as dominated.[5] The separation of these spheres has often been described as tragic, forcing both men and women into less than fully human, fully whole lives: men would be better off if they could integrate the human concerns of the domestic realm into their work lives and women would be better off if they were empowered to achieve on an equal basis with men in the workplace.[6]

Although social historians find that these categories emerged sharply in the nineteenth century, they are still with us in the conceptions that ordinary U.S. citizens today have of the world. In David Schneider's classic study of American kinship, conducted in the 1960s, the categories in which people explained their social life were the very ones being pushed to the center by the development of capitalism a century ago:

> The contrast between home and work brings out aspects which complete the picture of the distinctive features of kinship in American culture. This can best be understood in terms of the contrast between love and money which stand for home and work. Indeed, what one does at home, it is said, one does for love, not for money, while what one does at work one does strictly for money, not for love. Money is material, it is power, it is impersonal and universalistic, unqualified by considerations of sentiment and morality. Relations of work and money are temporary, transient, contingent. Love on the other hand is highly personal, and particularistic, and beset with considerations of sentiment and morality. Where love is spiritual, money is transient and contingent. And finally, it is personal considerations which are paramount in love—who the person is, not how well he performs, while with work and money it does not matter who he is, but only how well he performs his task.[7]

All these discussions in history and anthropology assume the existence of the individual, the person, one of our central cultural categories. Persons are composed of a body and a mind (some would add a soul) and are "relational monads" who are isolated from each other but necessarily in relationship with others.[8] Since it is possible in our way of thinking for some human beings to be less than full persons and since this condition seems to derive from the state of being dominated, it is quite likely that women are seen as less than fully persons.[9] It is highly suggestive that as U.S. law functions more and more as the main arbiter of relations among persons, women are consistently denied equal treatment with men.[10] This may be related to our notion that women are intrinsically closely involved with the family where so many "natural," "bodily" (and therefore lower) functions occur, whereas men are intrinsically closely involved with the world of work where (at least for some) "cultural," "mental," and therefore higher functions occur. It is no accident that "natural" facts about women, in the form of claims about biology, are often used to justify social stratification based on gender.

The other main way adult humans can be regarded as less than persons is on the basis of race, a social category masquerading as a natural one. Although no full-scale anthropological study has been done on how race as a social category operates in our society (an extraordinary omission that is in profound need of remedy), it seems clear that it cannot operate in just the same way that claims about gender do, so that black women are simply twice as oppressed as white women.[11] Claims have been made about the lesser intelligence of blacks or other minorities as they have been made about women. But those who make such claims (usually white males) can separate themselves from minority groups in a way they could never hope to separate themselves from women. Not only do most of them have a woman raising the kids at home, all of them surely believe that their children are genetically related—connected by shared biological substance—to their wives as well as themselves. Flaws in women might seem to have implications for their own families that flaws in other racial categories would not.

What happens when we follow the person as he or she moves out from the home to the workplace? In Marx's view, as Ollman has encapsulated it, humans at work in our kind of society experience alienation or separation of the parts of something that ought to be whole:

> An essential tie has been cut in the middle. Man is spoken of as being separated from his work (he plays no part in deciding what to do or how to do it)—a break between the individual and his life activity. Man is said to be separated from his own products (he has no control over what he makes or what becomes of it afterwards)—a break between the individual and the material world. He is also said to be separated from his fellow men (competition and class hostility has rendered most forms of cooperation impossible)—a break between man and man. . . . The whole has broken up into numerous parts whose interrelation in whole can no longer be ascertained. This is the essence of alienation, whether the part under examination is man, his activity, his product or his ideas. The same separation and distortion is evident in each.[12]

These seminal ideas have led many people to focus on what different forms alienation takes, how it can be detected, and how it varies in work experiences of different kinds.[13] In addition, many have tried to understand how alienation can be reduced. For example, Marx thought that such forms of separation and distortion were less prominent in prior stages of social evolution and that they could be eliminated in a future, ideal society.[14] Generally, it is assumed that separation of parts of the self (or of the self from the society) is debilitating and bad, and wholeness of the self is regenerating and good. Yet sometimes, as Sennet and Cobb have shown, separations can be adaptive in a particular setting. The working-class men they interviewed gained a measure of protection and ease from establishing separations within their selves: when external circumstances deny dignity to a person, letting only a portion of the self outside to be degraded is a form of self-preservation. "Dividing the self defends against the pain a person would otherwise feel, if he had to submit the whole of himself to a society which makes his position a vulnerable and anxiety-laden one."[15]

Although most empirical studies of alienation have involved male workers, some feminist theorists have tried to spell out how women's work is also alienated. Processes similar to those in the workplace affect women as mothers: whenever "women are forced into motherhood or prevented from becoming mothers, it is not they who decide how many children they bear."[16] They are constrained by circumstances outside their control. When they are actually bearing children, they "are viewed less as individuals than as the 'raw material' from which the 'product' is extracted. In these circumstances, the physician rather than the mother comes to be seen as having produced the baby." After the baby is born, the process of childrearing is affected in similar ways: "The child is a product which has to be produced according to exact specifications . . . mothers are ignorant of how to rear children and have to be instructed by experts. These experts, of course, are mostly male."[17]

Just as in the case of the worker in capitalism, women are separated from other persons around them: they are isolated from other mothers, at least in the middle class, by insular nuclear families. They are isolated from their children: "She sees the child as her product, as something that should improve her life and that often instead stands against her, as something of supreme value, that is held cheap by society. The social relations of contemporary motherhood make it impossible for her to see the child as a whole person, part of a larger community to which both mother and child belong." And they are isolated from the fathers of their children: "Rather than being parental co-workers with mothers, fathers often function as agents imposing the standards of the larger society."[18]

Moving down a level, from the person to the relationships among the parts of a person, we also find many descriptions of fragmentation and loss of wholeness. One such description is Marx's notion of alienation caused by separation of the worker from his or her work. Alienated labor stands in the way of what he thought labor could ideally be: the quintessential human activity involving a unity of purpose and planning toward some end significant to the worker.[19]

In addition, many elements of modern medical science have been held to contribute to a fragmentation of the unity of the person. When science treats the person as a machine and assumes the body can be fixed by mechanical manipulations, it ignores, and it encourages us to ignore, other aspects of our selves, such as our emotions or our relations with other people.[20] Recent technological developments have allowed this tendency to progress very far. Parts of our bodies can now be moved from person to person; their purchase and sale can even be contemplated.[21] The body as a machine without a mind or soul has become almost familiar, but the body without the integrity of even its parts will necessarily lead to many readjustments in our conceptions of the self, and the shape that will emerge is far from clear. Transplant patients, particularly those with new hearts, an organ laden with symbolic significance as the deepest seat of the emotional self, are often asked what the experience is like. "'It's sort of a strange concept to understand,' [a transplant patient] reflects, 'that my heart was taken out and someone else's put in. But I've never had any problems with the idea. It's only been a few months but I've gotten so used to the fact that I had a heart transplant that it doesn't faze me.'"[22]

And on the other side, what about the person who is declared dead but whose body parts are transplanted into a living person? A mother "finds some consolation in knowing [her son's] organs are giving life to someone else. One of his kidneys went to a 16-year-old girl in Washington, the other to a 28-year-old Maryland man who had just gotten married. A cornea went to a 34-year-old Baltimore woman. The Black family learned these few details about the recipients of their son's organs on the day of his funeral. 'It has helped,' Mrs. Black says, 'it really has helped. Because of the transplants, I feel that somehow Glenn is still walking around somewhere.'"[23]

Similar processes, of course, affect the whole course of reproduction. Human eggs, sperm, and embryos can now be moved from body to body or out of and back into the same female body. The organic unity of fetus and mother can no longer be assumed, and all these newly fragmented parts can now be subjected to market forces, ordered, produced, bought, and sold.[24]

Foucault begins his *Discipline and Punish* with a riveting description of an eighteenth-century public spectacle of torture and dismemberment. He comments that the decline of such spectacles "marks a slackening of the hold on the body." In contemporary society, if it is necessary for the law to manipulate the body of a convict, make it work or imprison it, "it is in order to deprive the individual of a liberty that is regarded both as a right and as property . . . Physical pain, the pain of the body itself, is no longer the constituent element of the penalty."[25] Foucault is surely right in pointing to the different role of physical pain in the two eras, as epitomized by our use of drugs to prevent pain or anxiety during an execution. But dismemberment is with us still, and the "hold on the body" has not so much slackened as it has moved from the law to science.

Women, it has been argued, suffer the alienation of parts of the self much more acutely than men. For one thing, becoming sexually female entails inner fragmentation of the self. A woman must become only a physical body in order to be sexual:

> A truly "feminine" woman, then, has been seduced by a variety of cultural agencies into being a body not only for another, but for herself, as well. But when this happens, she may well experience what is in effect a taboo on the development of her other human capacities. In our society, for example, the cultivation of intellect has made a woman not more but less sexually alluring.[26]

Beyond this, the body she becomes is itself an object to her:

> Knowing that she is to be subjected to the cold appraisal of the male connoisseur and that her life prospects may depend on how she is seen, a woman learns to appraise herself first. The sexual objectification of woman produces a duality in feminine consciousness . . . What occurs is not just the splitting of a person into mind and body but the splitting of the self into a number of *personae,* some who witness and some who are witnessed.[27]

Finally, women are not only fragmented into body parts by the practices of scientific medicine, as men are; they are also profoundly alienated from science itself. They are less involved in the production of science as a form of cultural activity because of "a network of interactions between gender development, a belief system that equates objectivity with masculinity, and a set of cultural values that simultaneously (and conjointly) elevates what is defined as scientific and what is defined as masculine."[28] In addition, the content of science presents "a male-biased model of human nature and social reality."[29]

> We find that the attributes of science are the attributes of males; the objectivity said to be characteristic of the production of scientific knowledge is specifically identified as a male way of relating to the world. Science is cold, hard, impersonal, "objective"; women, by contrast, are warm, soft, emotional, "subjective."[30]

This depiction of modern consciousness with its fragmentations and alienations is based on a number of sources: testimony gathered from ordinary workers and citizens; philosophical analysis about the nature of human life and work; and the personal introspection and group discussion of concerned theorists. My task in this book is to confront the claims about our cultural representations that I have just described with the empirical data I have gathered. I will ask how women in different social and economic positions see themselves and how they see their society. In particular, I will ask what picture of reality they convey when they are asked to talk not about their families, spouses, and children (when they seem very likely to simply reproduce a version of dominant cultural ideology) but about themselves, through the medium of events which only women experience and which perhaps for that reason are rarely spoken of—menstruation, childbirth, and menopause.

The answers will turn out to be complex, but the guiding questions are simple. The depiction of modern consciousness leads to the conclusion that women's lives are especially degraded, fragmented, and impoverished. At least this is one reading of things. The imperative question is, How do women react to their circumstances? Do they describe their existence in the terms used by medical science and dominant society? If so, do they find them acceptable and unquestionable or lamentable but not changeable facts of life? Or do they find them outrageous and intolerable? Is there anything approaching a woman's alternative vision of modern existence, a woman's ideology, or are there many alternative visions, refractions of women's many different places in the social order? Or are women's views simply reflections of the dominant cultural view, diminishing as it may be of women's experience?

My intent has been to base this analysis on data that reflect the experience of women in a wide range of economic and social positions in our society and data that are capable of carrying the rich inflections with which people often invest their lives with meaning. If women are one of those "muted" groups,[31] subject to a relatively great degree of oppression, such that they may not always know their oppression, object to it, or resist it, then we must have extremely sensitive ways of looking for evidence of women's consciousness of their situation and for a wide variety of forms of objection or resistance.

To deal with the way women relate to dominant cultural categories through the medium of conversations about their bodily processes, I must make the cultural analysis of scientific representations of women's bodies primary. Although we tend to think of science as outside culture because it seeks the truth about nature, I assert that it is in fact more like a hegemonic system. Hegemony, a term used by Gramsci, means

> the permeation throughout civil society . . . of an entire system of values, attitudes, beliefs, morality, etc. that is in one way or another supportive of the established order and the class interests that dominate it . . . to the extent that this prevailing consciousness is internalized by the broad masses, it becomes part of "common sense" . . . For hegemony to assert itself successfully in any society, therefore, it must operate in a dualistic manner: as a "general conception of life" for the masses and as a "scholastic programme."[32]

My task in Part Two will be to tease out the "general conception of life" that is entailed in dominant scientific ideas about women's bodies. My task in Part Three is to see, as women talk about major physical events they experience from menstruation to menopause, whether women are aware of these scientific ideas as well as other ideas about gender in this society, and if so whether they accept them or resist them. We will be able to see in what senses and to what extent science has become women's common sense. In Part Four I deal with the relation between class position and consciousness, asking whether women are more or less mystified about the nature of their oppression at lower levels in the hierarchy of class and race. I also address the related question of whether women have an alternative consciousness based on their different life experiences. The claim for such a consciousness has been made in theoretical terms, but this study allows us to see for the first time whether an alternative consciousness is in fact present in the way ordinary women think about and act in the world.

Science as a Cultural System

3

Medical Metaphors of Women's Bodies: Menstruation and Menopause

Lavoisier makes experiments with substances in his laboratory and now he concluded that this and that takes place when there is burning. He does not say that it might happen otherwise another time. He has got hold of a definite world-picture—not of course one that he invented: he learned it as a child. I say world-picture and not hypothesis, because it is the matter-of-course foundation for his research and as such also goes unmentioned.
—Ludwig Wittgenstein
On Certainty

It is difficult to see how our current scientific ideas are infused by cultural assumptions; it is easier to see how scientific ideas from the past, ideas that now seem wrong or too simple, might have been affected by cultural ideas of an earlier time. To lay the groundwork for a look at contemporary scientific views of menstruation and menopause, I begin with the past.

It was an accepted notion in medical literature from the ancient Greeks until the late eighteenth century that male and female bodies were structurally similar. As Nemesius, bishop of Emesa, Syria, in the fourth century, put it, "women have the same genitals as men, except that theirs are inside the body and not outside it." Although increasingly detailed anatomical understanding (such as the discovery of the nature of the ovaries in the last half of the seventeenth century) changed the details, medical scholars from Galen in second-century Greece to Harvey in seventeenth-century Britain all assumed that women's internal organs were structurally analogous to men's external ones.[1] (See Figures 1–4.)

Although the genders were structurally similar, they were not

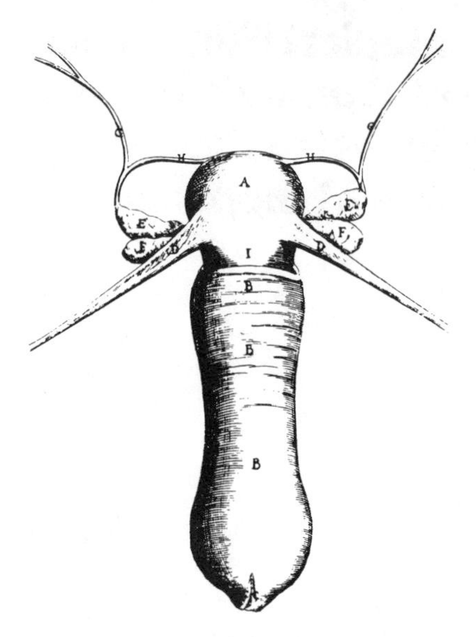

Fig. 1 Vidius' depiction of the uterus and vagina as analogous to the penis and scrotum. (Vidius 1611, Vol. 3. Photo taken from Weindler 1908:140.)

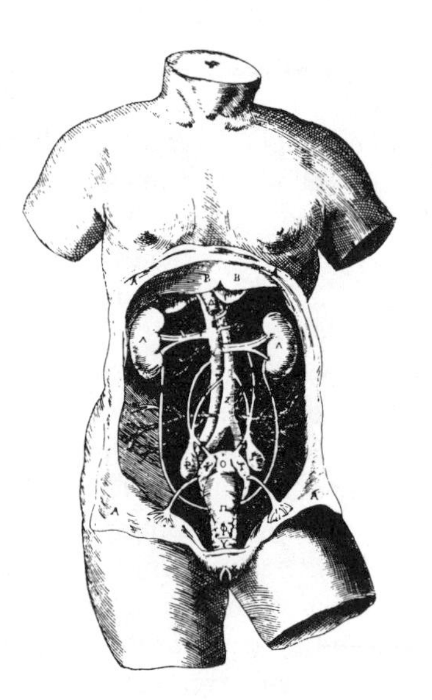

Fig. 2 Vidius' illustration of how the female organs are situated inside the body. (Vidius 1611, Vol. 3. Photo taken from Weindler 1908:139.)

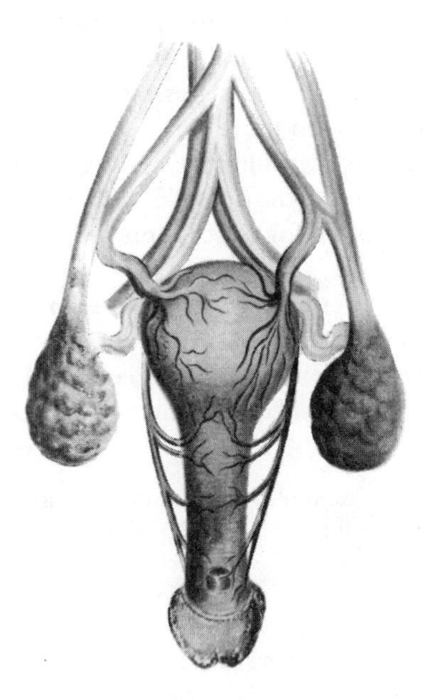

Fig. 3 Georg Bartisch's illustration of phallus-like female reproductive organs. (Attributed by Weindler 1908:141 to Bartisch's *Kunstbuche,* 1575 [MS Dresdens. C. 291]. Photo taken from Weindler 1908, fig. 104b, p. 144.)

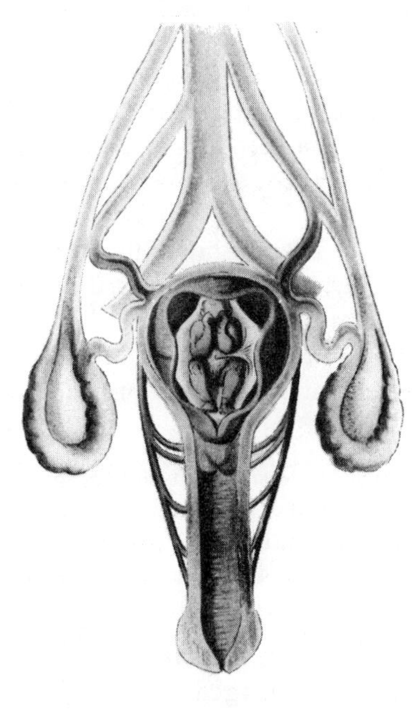

Fig. 4 Bartisch's cross-section of the female organs, showing a fetus inside the uterus. (Attributed by Weindler to Bartisch's *Kunstbuche,* 1575 [MS Dresdens. C. 291]. Photo taken from Weindler 1908, fig. 104b, p. 144.)

equal. For one thing, what could be seen of men's bodies was assumed as the pattern for what could not be seen of women's. For another, just as humans as a species possessed more "heat" than other animals, and hence were considered more perfect, so men possessed more "heat" than women and hence were considered more perfect. The relative coolness of the female prevented her reproductive organs from extruding outside the body but, happily for the species, kept them inside where they provided a protected place for conception and gestation.[2]

During the centuries when male and female bodies were seen as composed of analogous structures, a connected set of metaphors was used to convey how the parts of male and female bodies functioned. These metaphors were dominant in classical medicine and continued to operate through the nineteenth century:

> The body was seen, metaphorically, as a system of dynamic interactions with its environment. Health or disease resulted from a cumulative interaction between constitutional endowment and environmental circumstance. One could not well live without food and air and water; one had to live in a particular climate, subject one's body to a particular style of life and work. Each of these factors implied a necessary and continuing physiological adjustment. The body was always in a state of becoming—and thus always in jeopardy.[3]

Two subsidiary assumptions governed this interaction: first, that "every part of the body was related inevitably and inextricably with every other" and, second, that "the body was seen as a system of intake and outgo—a system which had, necessarily, to remain in balance if the individual were to remain healthy."[4]

Given these assumptions, changes in the relationship of body functions occurred constantly throughout life, though more acutely at some times than at others. In Edward Tilt's influential mid-nineteenth-century account, for example, after the menopause blood that once flowed out of the body as menstruation was then turned into fat:

> Fat accumulates in women after the change of life, as it accumulates in animals from whom the ovaries have been removed. The withdrawal of the sexual stimulus from the ganglionic nervous system, enables it to turn into fat and self-aggrandisement that blood which might otherwise have perpetuated the race.[5]

During the transition to menopause, or the "dodging time," the blood could not be turned into fat, so it was either discharged as hemorrhage or through other compensating mechanisms, the most important of which was "the flush":

> As for thirty-two years it had been habitual for women to lose about 3 oz. of blood every month, so it would have been indeed singular, if there did not exist some well-continued compensating discharges acting as waste-gates to protect the system, until health could be permanently re-established by striking new balances in the allotment of blood to the various parts . . . The flushes determine the perspirations. Both evidence a strong effect of conservative power, and as they constitute the most important and habitual safety-valve of the system at the change of life, it is worth while studying them.[6]

In this account, compensating mechanisms like the "flush" are seen as having the positive function of keeping intake and outgo in balance.

These balancing acts had exact analogues in men. In Hippocrates' view of purification, one that was still current in the seventeenth century,

> women were of a colder and less active disposition than men, so that while men could sweat in order to remove the impurities from their blood, the colder dispositions of women did not allow them to be purified in that way. Females menstruated to rid their bodies of impurities.[7]

Or in another view, expounded by Galen in the second century and still accepted into the eighteenth century, menstruation was the shedding of an excess of blood, a plethora.[8] But what women did through menstruation men could do in other ways, such as by having blood let.[9] In either view of the mechanism of menstruation, the process itself not only had analogues in men, it was seen as inherently health-maintaining. Menstrual blood, to be sure, was often seen as foul and unclean,[10] but the process of excreting it was not intrinsically pathological. In fact, failure to excrete was taken as a sign of disease, and a great variety of remedies existed even into the nineteenth century specifically to reestablish menstrual flow if it stopped.[11]

By 1800, according to Laqueur's important recent study, this long-established tradition that saw male and female bodies as similar both in structure and in function began to come "under devastating attack. Writers of all sorts were determined to base what they insisted were fundamental differences between male and female sexuality, and thus between man and woman, on discoverable biological distinctions."[12] Laqueur argues that this attempt to ground differences between the genders in biology grew out of the crumbling of old ideas about the existing order of politics and society as laid down by the order of nature. In the old ideas, men dominated the public world and the world of morality and order by virtue of their greater perfection, a result of their excess heat. Men and women were arranged in a hierarchy in which they differed by degree of heat. They were not different in kind.[13]

The new liberal claims of Hobbes and Locke in the seventeenth century and the French Revolution were factors that led to a loss of certainty that the social order could be grounded in the natural order. If the social order were merely convention, it could not provide a secure enough basis to hold women and men in their places. But after 1800 the social and biological sciences were brought to the rescue of male superiority. "Scientists in areas as diverse as zoology, embryology, physiology, heredity, anthropology, and psychology had little difficulty in proving that the pattern of male-female relations that characterized the English middle classes was natural, inevitable, and progressive."[14]

The assertion was that men's and women's social roles themselves were grounded in nature, by virtue of the dictates of their bodies. In the words of one nineteenth-century theorist, "the attempt to alter the present relations of the sexes is not a rebellion against some arbitrary law instituted by a despot or a majority—not an attempt to break the yoke of a mere convention; it is a struggle against Nature; a war undertaken to reverse the very conditions under which not man alone, but all mammalian species have reached their present development."[15] The doctrine of the two spheres discussed in the last chapter—men as workers in the public, wage-earning sphere outside the home and women (except for the lower classes) as wives and mothers in the private, domestic sphere of kinship and morality inside the home—replaced the old hierarchy based on body heat.

During the latter part of the nineteenth century, new metaphors that posited fundamental differences between the sexes began to appear. One nineteenth-century biologist, Patrick Geddes, perceived two opposite kinds of processes at the level of the cell: "upbuilding, constructive, synthetic prosesses," summed up as anabolism, and a "disruptive, descending series of chemical changes," summed up as katabolism.[16] The relationship between the two processes was described in frankly economic terms:

> . . . The processes of income and expenditure must balance, but only to the usual extent, that expenditure must not altogether outrun income, else

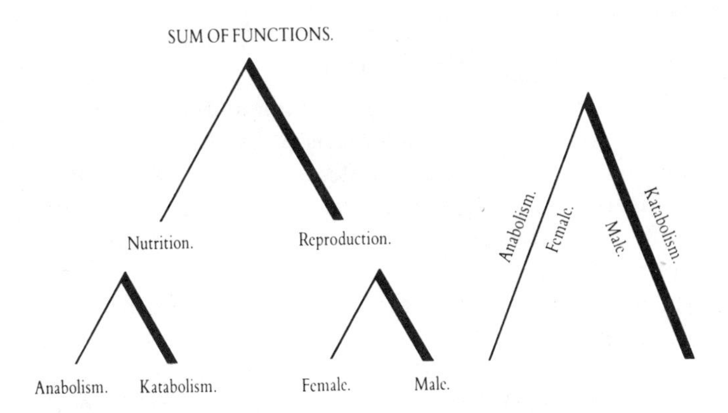

Fig. 5 An illustration accompanying the late nineteenth-century biologist Geddes's account of what he saw as radical physiological distinctions between males and females, the male dominated by active, energetic katabolic functions and the female by passive, conservative anabolic functions. (Geddes 1890:213.)

the cell's capital of living matter will be lost,—a fate which is often not successfully avoided . . . Just as our expenditure and income should balance at the year's end, but may vastly outstrip each other at particular times, so it is with the cell of the body. Income too may continuously preponderate, and we increase in wealth, or similarly, in weight, or in anabolism. Conversely, expenditure may predominate, but business may be prosecuted at a loss; and similarly, we may live on for a while with loss of weight, or in katabolism. This losing game of life is what we call a katabolic habit.[17]

Geddes saw these processes not only at the level of the cell, but also at the level of entire organisms. In the human species, as well as in almost all higher animals, females were predominantly anabolic, males katabolic. (See Figure 5.) Although in the terms of his saving-spending metaphor it is not at all clear whether katabolism would be an asset, when Geddes presents male-female differences, there is no doubt which he thought preferable:

It is generally true that the males are more active, energetic, eager, passionate, and variable; the females more passive, conservative, sluggish, and stable . . . The more active males, with a consequently wider range of experience, may have bigger brains and more intelligence; but the females, especially as mothers, have indubitably a larger and more habitual share of the altruistic emotions. The males being usually stronger, have greater independence and courage; the females excel in constancy of affection and in sympathy.[18]

In Geddes, the doctrine of separate spheres was laid on a foundation of separate and fundamentally different biology in men and women, at the level of the cell. One of the striking contradictions in his account is that he did not carry over the implications of his economic metaphors to his discussion of male-female differences. If he had, females might have come off as wisely conserving their energy and never spending beyond their means, males as in the "losing game of life," letting expenditures outrun income.

Geddes may have failed to draw the logical conclusions from his metaphor, but we have to acknowledge that metaphors were never meant to be logical. Other nineteenth-century writers developed metaphors in exactly opposite directions: women spent and men saved. The Rev. John Todd saw women as voracious spenders in the marketplace, and so consumers of all that a man could earn. If unchecked, a woman would ruin a man, by her own extravagant spending, by her demands on him to spend, or, in another realm, by her excessive demands on him for sex. Losing too much sperm meant losing that which sperm was believed to manufacture: a man's lifeblood.[19]

Todd and Geddes were not alone in the nineteenth century in using images of business loss and gain to describe physiological processes. Susan Sontag has suggested that nineteenth-century fantasies about disease, especially tuberculosis, "echo the attitudes of early capitalist accumulation. One has a limited amount of energy, which must be properly spent . . . Energy, like savings, can be depleted, can run out or be used up, through reckless expenditure. The body will start 'consuming' itself, the patient will 'waste away.'"[20]

Despite the variety of ways that spending-saving metaphors could be related to gender, the radical difference between these metaphors and the earlier intake-outgo metaphor is key. Whereas in the earlier model, male and female ways of secreting were not only analogous but desirable, now the way became open to denigrate, as Geddes overtly did, functions that for the first time were seen as uniquely female, without analogue in males. For our purposes, what happened to accounts of menstruation is most interesting: by the nineteenth century, the process itself was seen as soundly pathological. In Geddes' terms,

> it yet evidently lies on the borders of pathological change, as is evidenced not only by the pain which so frequently accompanies it, and the local and constitutional disorders which so frequently arise in this connection, but by the general systemic disturbance and local histological changes of which the discharge is merely the outward expression and result.[21]

Whereas in earlier accounts the blood itself may have been considered impure, now the process itself is seen as a disorder.

Nineteenth-century writers were extremely prone to stress the debilitating nature of menstruation and its adverse impact on the lives and activities of women.[22] Medical images of menstruation as pathological were remarkably vivid by the end of the century. For Walter Heape, the militant antisuffragist and Cambridge zoologist, in menstruation the entire epithelium was torn away,

> leaving behind a ragged wreck of tissue, torn glands, ruptured vessels, jagged edges of stroma, and masses of blood corpuscles, which it would seem hardly possible to heal satisfactorily without the aid of surgical treatment.[23]

A few years later, Havelock Ellis could see women as being "periodically wounded" in their most sensitive spot and "emphasize the fact that even in the healthiest woman, a worm however harmless and unperceived, gnaws periodically at the roots of life."[24]

If menstruation was consistently seen as pathological, menopause, another function which by this time was regarded as without analogue in men, often was too: many nineteenth-century medical accounts of menopause saw it as a crisis likely to bring on an increase of disease.[25] Sometimes the metaphor of the body as a small business that is either winning or losing was applied to menopause too. A late-nineteenth-century account specifically argued against Tilt's earlier adjustment model: "When the period of fruitfulness is ended the activity of the tissues has reached its culmination, the secreting power of the glandular organs begins to diminish, the epithelium becomes less sensitive and less susceptible to infectious influences, and atrophy and degeneration take the place of the active up-building processes."[26] But there were other sides to the picture. Most practitioners felt the "climacteric disease," a more general disease of old age, was far worse for men than for women.[27] And some regarded the period after menopause far more positively than it is being seen medically in our century, as the "'Indian summer' of a woman's life—a period of increased vigor, optimism, and even of physical beauty.'"[28]

Perhaps the nineteenth century's concern with conserving energy and limiting expenditure can help account for the seeming anomaly of at least some positive medical views of menopause and the climacteric. As an early-twentieth-century popular health account put it,

> [Menopause] is merely a conservative process of nature to provide for a higher and more stable phase of existence, an economic lopping off of a function no longer needed, preparing the individual for different forms of

activity, but is in no sense pathologic. It is not sexual or physical decrepitude, but belongs to the age of invigoration, marking the fullness of the bodily and mental powers.[29]

Those few writers who saw menopause as an "economic" physiological function might have drawn very positive conclusions from Geddes' description of females as anabolic, stressing their "thriftiness" instead of their passivity, their "growing bank accounts" instead of their sluggishness.

If the shift from the body as an intake-outgo system to the body as a small business trying to spend, save, or balance its accounts is a radical one, with deep importance for medical models of female bodies, so too is another shift that began in the twentieth century with the development of scientific medicine. One of the early-twentieth-century engineers of our system of scientific medicine, Frederick T. Gates, who advised John D. Rockefeller on how to use his philanthropies to aid scientific medicine, developed a series of interrelated metaphors to explain the scientific view of how the body works:

It is interesting to note the striking comparisons between the human body and the safety and hygienic appliances of a great city. Just as in the streets of a great city we have "white angels" posted everywhere to gather up poisonous materials from the streets, so in the great streets and avenues of the body, namely the arteries and the blood vessels, there are brigades of corpuscles, white in color like the "white angels," whose function it is to gather up into sacks, formed by their own bodies, and disinfect or eliminate all poisonous substances found in the blood. The body has a network of insulated nerves, like telephone wires, which transmit instantaneous alarms at every point of danger. The body is furnished with the most elaborate police system, with hundreds of police stations to which the criminal elements are carried by the police and jailed. I refer to the great numbers of sanitary glands, skilfully placed at points where vicious germs find entrance, especially about the mouth and throat. The body has a most complete and elaborate sewer system. There are wonderful laboratories placed at convenient points for a subtle brewing of skillful medicines . . . The fact is that the human body is made up of an infinite number of microscopic cells. Each one of these cells is a small chemical laboratory, into which its own appropriate raw material is constantly being introduced, the processes of chemical separation and combination are constantly taking place automatically, and its own appropriate finished product being necessary for the life and health of the body. Not only is this so, but the great organs of the body like the liver, stomach, pancreas, kidneys, gall bladder are great local manufacturing centers, formed of groups of cells in infinite numbers, manufacturing the same sorts of products, just as industries of the same kind are often grouped in specific districts.[30]

Although such a full-blown description of the body as a model of an industrial society is not often found in contemporary accounts of physiology, elements of the images that occurred to Gates are commonplace. In recent years, the "imagery of the biochemistry of the cell [has] been that of the factory, where functions [are] specialized for the conversion of energy into particular products and which [has] its own part to play in the economy of the organism as a whole."[31] There is no doubt that the basic image of cells as factories is carried into popular imagination, and not only through college textbooks: the illustration from *Time* magazine shown in Figure 6 depicts cells explicitly as factories (and AIDS virus cells as manufacturing armored tanks!).

Still more recently, economic functions of greater complexity have been added: ATP is seen as the body's "energy currency": "Produced in particular cellular regions, it [is] placed in an 'energy bank' in which it [is] maintained in two forms, those of 'current account' and 'deposit account.' Ultimately, the cell's and the body's energy books must balance by an appropriate mix of monetary and fiscal policies."[32] Here we have not just the simpler nineteenth-century saving and spending, but two distinct forms of money in the bank, presumably invested at different levels of profit.

Development of the new molecular biology brought additional metaphors based on information science, management, and control. In this model, flow of information between DNA and RNA leads to the production of protein.[33] Molecular biologists conceive of the cell as "an assembly line factory in which the DNA blueprints are interpreted and raw materials fabricated to produce the protein end products in response to a series of regulated requirements."[34] The cell is still seen as a factory, but, compared to Gates' description, there is enormous elaboration of the flow of information from one "department" of the body to another and exaggeration of the amount of control exerted by the center. For example, from a college physiology text:

> All the systems of the body, if they are to function effectively, must be subjected to some form of control . . . The precise control of body function is brought about by means of the operation of the nervous system and of the hormonal or endocrine system . . . The most important thing to note about any control system is that before it can control anything it must be supplied

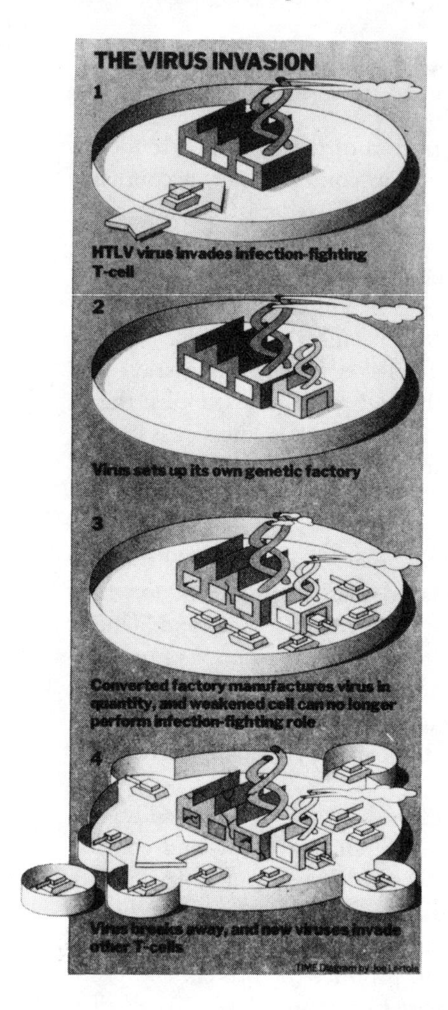

Fig. 6 A contemporary image of cells as factories. (*Time* magazine, 30 April 1984:67. Copyright 1984 by Time, Inc. All rights reserved. Reprinted by permission from TIME.)

with information . . . Therefore the first essential in any control system is an adequate system of collecting information about the state of the body . . . Once the CNS [central nervous system] knows what is happening, it must then have a means for rectifying the situation if something is going wrong. There are two available methods for doing this, by using nerve fibres and by using hormones. The motor nerve fibres . . . carry instructions from the CNS to the muscles and glands throughout the body . . . As far as hormones are concerned the brain acts via the pituitary gland . . . the pituitary secretes a large number of hormones . . . the rate of secretion of each one of these is under the direct control of the brain.[35]

The illustration in Figure 7 reiterates this account vividly: there is a "co-ordinating centre" which transmits messages to and receives messages from peripheral parts, for the purpose of integration and control. Although there is increasing attention to describing physiological processes as positive and negative feedback loops so that like a

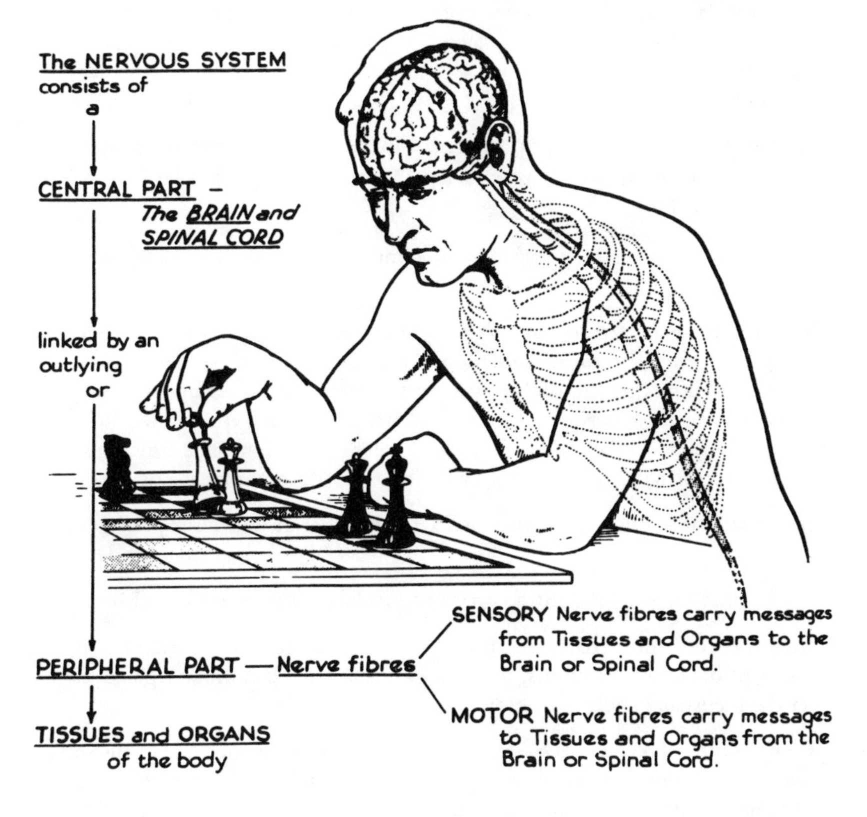

Fig. 7 An image from a text for premedical students showing the brain as a coordinating center transmitting messages to and receiving them from outlying parts. (McNaught and Callander 1983:204. Reprinted by permission of Churchill Livingstone.)

thermostat system no single element has preeminent control over any other, most descriptions of specific processes give preeminent control to the brain, as we will see below.

Metaphors in Descriptions of Female Reproduction

In overall descriptions of female reproduction, the dominant image is that of a signaling system. Lein, in a textbook designed for junior colleges, spells it out in detail:

> Hormones are chemical signals to which distant tissues or organs are able to respond. Whereas the nervous system has characteristics in common with a telephone network, the endocrine glands perform in a manner somewhat analogous to radio transmission. A radio transmitter may blanket an entire region with its signal, but a response occurs only if a radio receiver is turned on and tuned to the proper frequency . . . the radio receiver in biological systems is a tissue whose cells possess active receptor sites for a particular hormone or hormones.[36]

The signal-response metaphor is found almost universally in current texts for premedical and medical students (emphasis in the following quotes is added):

> The hypothalamus *receives signals* from almost all possible sources in the nervous system.[37]

> The endometrium *responds directly* to stimulation or withdrawal of estrogen and progesterone. In turn, regulation of the secretion of these steroids involves a well-integrated, highly structured series of activities by the hypothalamus and the anterior lobe of the pituitary. Although the ovaries do not function autonomously, they *influence,* through *feedback* mechanisms, the level of performance *programmed* by the hypothalamic-pituitary axis.[38]

> As a result of strong stimulation of FSH, a number of follicles *respond* with growth.[39]

And the same idea is found, more obviously, in popular health books:

> Each month from menarch on, [the hypothalamus] acts as elegant interpreter of the body's rhythms, *transmitting messages* to the pituitary gland that set the menstrual cycle in motion.[40]

> Each month, *in response to a message* from the pituitary gland, one of the unripe egg cells develops inside a tiny microscopic ring of cells, which gradually increases to form a little balloon or cyst called the Graafian follicle.[41]

Although most accounts stress signals or stimuli traveling in a "loop" from hypothalamus to pituitary to ovary and back again, carrying positive or negative feedback, one element in the loop, the hypothalamus, a part of the brain, is often seen as predominant. Just as in the general model of the central nervous system shown in Figure 7, the female brain-hormone-ovary system is usually described not as a feedback loop like a thermostat system, but as a hierarchy, in which the "directions" or "orders" of one element dominate (emphasis in the following quotes from medical texts is added):

Both positive and negative feedback control must be invoked, together with *superimposition* of control by the CNS through neurotransmitters released into the hypophyseal portal circulation.[42]

Almost all secretion by the pituitary is *controlled* by either hormonal or nervous signals from the hypothalamus.[43]

The hypothalamus is a collecting center for information concerned with the internal well-being of the body, and in turn much of this information is used *to control* secretions of the many globally important pituitary hormones.[44]

As Lein puts it into ordinary language, "The cerebrum, that part of the brain that provides awareness and mood, can play a significant role in the control of the menstrual cycle. As explained before, it seems evident that these higher regions of the brain exert their influence by modifying the actions of the hypothalamus. So even though the hypothalamus is a kind of master gland dominating the anterior pituitary, and through it the ovaries also, it does not act with complete independence or without influence from outside itself . . . there are also pathways of control from the higher centers of the brain."[45]

So this is a communication system organized hierarchically, not a committee reaching decisions by mutual influence.[46] The hierarchical nature of the organization is reflected in some popular literature meant to explain the nature of menstruation simply: "From first menstrual cycle to menopause, the hypothalamus acts as the conductor of a highly trained orchestra. Once its baton signals the downbeat to the pituitary, the hypothalamus-pituitary-ovarian axis is united in purpose and begins to play its symphonic message, preparing a woman's body for conception and child-bearing." Carrying the metaphor further, the follicles vie with each other for the role of producing the egg like violinists trying for the position of concertmaster; a burst of estrogen is emitted from the follicle like a "clap of tympani."[47]

The basic images chosen here—an information-transmitting system with a hierarchical structure—have an obvious relation to the dominant form of organization in our society.[48] What I want to show is how this set of metaphors, once chosen as the basis for the description of physiological events, has profound implications for the way in which a change in the basic organization of the system will be perceived. In terms of female reproduction, this basic change is of course menopause. Many criticisms have been made of the medical propensity to see menopause as a pathological state.[49] I would like to suggest that the tenacity of this view comes not only from the negative stereotypes associated with aging women in our society, but as a logical outgrowth of seeing the body as a hierarchical information-processing system in the first place. (Another part of the reason menopause is seen so negatively is related to metaphors of production, which we discuss later in this chapter.)

What is the language in which menopause is described? In menopause, according to a college text, the ovaries become "unresponsive" to stimulation from the gonadotropins, to which they used to respond. As a result the ovaries "regress." On the other end of the cycle, the hypothalamus has gotten estrogen "addiction" from all those years of menstruating. As a result of the "withdrawal" of estrogen at menopause, the hypothalamus begins to give "inappropriate orders."[50] In a more popular account, "the pituitary gland during the change of life becomes disturbed when the ovaries fail to respond to its secretions, which tends to affect its control over other glands. This results in a temporary imbalance existing among all the endocrine glands of the body, which could very well lead to disturbances that may involve a person's nervous system."[51]

In both medical texts and popular books, what is being described is the breakdown of a system of authority. The cause of ovarian "decline" is the "decreasing ability of the aging ovaries to respond to pituitary gonadotropins."[52] At every point in this system, functions "fail" and falter. Follicles "fail to muster the strength" to reach ovulation.[53] As functions fail, so do the members of the system decline: "breasts and genital organs gradually atrophy,"[54] "wither,"[55] and become "senile."[56] Diminished, atrophied relics of their former vigorous, functioning selves, the "senile ovaries" are an example of the vivid imagery brought to this process. A text whose detailed illustrations make it a primary resource for medical students despite its early date describes the ovaries this way:

> the *senile ovary* is a shrunken and puckered organ, containing few if any follicles, and made up for the most part of old corpora albincantia and corpora atretica, the bleached and functionless remainders of corpora lutia and follicles embedded in a dense connective tissue stroma.[57]

MENOPAUSE

Between the ages of 42 and 50 years OVARIAN tissue gradually ceases to respond to stimulation by ANTERIOR PITUITARY GONADOTROPHIC HORMONES.

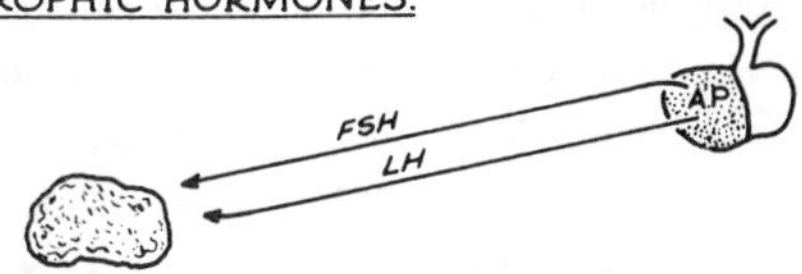

OVARIAN CYCLE becomes irregular and finally ceases → Ovary becomes small and fibrosed and no longer produces ripe Ova.
OESTROGEN and PROGESTERONE levels in Blood stream fall.
TISSUES of the body — begin to show changes which mark the end of REPRODUCTIVE LIFE.

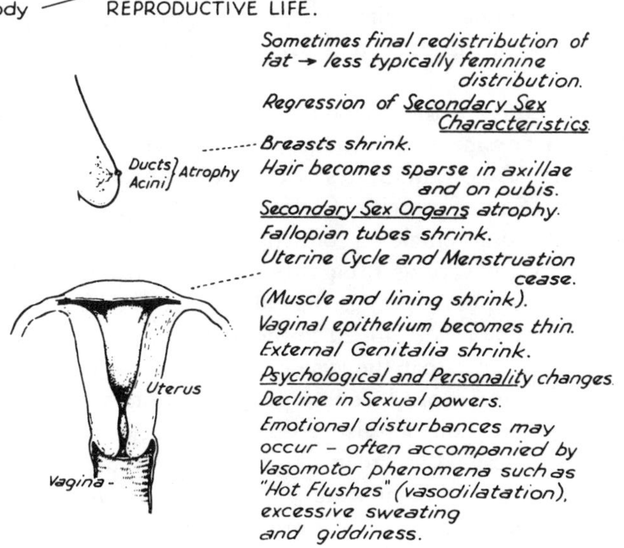

After the MENOPAUSE a woman is usually unable to bear children.

Fig. 8 A summary diagram from a text for premedical students that emphasizes menopause as a process of breakdown, failure, and decline. (McNaught and Callander 1983:200. Reprinted by permission of Churchill Livingstone.)

The illustration in Figure 8 summarizes the whole picture: ovaries cease to respond and fail to produce. Everywhere else there is regression, decline, atrophy, shrinkage, and disturbance.

The key to the problem connoted by these descriptions is functionlessness. Susan Sontag has written of our obsessive fear of cancer, a disease that we see as entailing a nightmare of excessive growth and rampant production. These images frighten us in part because in our stage of advanced capitalism, they are close to a reality we find difficult to see clearly: broken-down hierarchy and organization members who no longer play their designated parts represent nightmare images for us. To anticipate a later chapter, one woman I talked to said her doctor gave her two choices for treatment of her menopause: she could take estrogen and get cancer or she could not take it and have her bones dissolve. Like this woman, our imagery of the body as a hierarchical organization gives us no good choice when the basis of the organization seems to us to have changed drastically. We are left with breakdown, decay, and atrophy. Bad as they are, these might be preferable to continued activity, which because it is not properly hierarchically controlled, leads to chaos, unmanaged growth, and disaster.

But let us return to the metaphor of the factory producing substances, which dominates the imagery used to describe cells. At the cellular level DNA communicates with RNA, all for the purpose of the cell's production of proteins. In a similar way, the system of communication involving female reproduction is thought to be geared toward production of various things. In the next chapter we look in detail at images of production as they affect labor and birth. For the present this discussion is confined to the normal process of the menstrual cycle. It is clear that the system is thought to produce many good things: the ovaries produce estrogen, the pituitary produces FSH and LH, and so on. Follicles also produce eggs in a sense, although this is usually described as "maturing" them since the entire set of eggs a woman has for her lifetime is known to be present at birth. Beyond all this the system is seen as organized for a single preeminent purpose: "transport" of the egg along its journey from the ovary to the uterus[58] and preparation of an appropriate place for the egg to grow if it is fertilized. In a chapter titled "Prepregnancy Reproductive Functions of the Female, and the Female Hormones," Guyton puts it all together: "Female reproductive functions can be divided into two major phases: first, preparation of the female body for conception and gestation, and second, the period of gestation itself."[59] This view may seem commonsensical and entirely justified by the evolutionary development of the species, with its need for reproduction to ensure survival.

Yet I suggest that assuming this view of the purpose for the process slants our description and understanding of the female cycle unnecessarily. Let us look at how medical textbooks describe menstruation.

They see the action of progesterone and estrogen on the lining of the uterus as "ideally suited to provide a hospitable environment for implantation and survival of the embryo"[60] or as intended to lead to "the monthly renewal of the tissue that will cradle [the ovum]."[61] As Guyton summarizes, "The whole purpose of all these endometrial changes is to produce a highly secretory endometrium containing large amounts of stored nutrients that can provide appropriate conditions for implantation of a fertilized ovum during the latter half of the monthly cycle."[62] Given this teleological interpretation of the purpose of the increased amount of endometrial tissue, it should be no surprise that when a fertilized egg does not implant, these texts describe the next event in very negative terms. The fall in blood progesterone and estrogen "deprives" the "highly developed endometrial lining of its hormonal support," "constriction" of blood vessels leads to a "diminished" supply of oxygen and nutrients, and finally "disintegration starts, the entire lining begins to slough, and the menstrual flow begins." Blood vessels in the endometrium "hemorrhage" and the menstrual flow "consists of this blood mixed with endometrial debris."[63] The "loss" of hormonal stimulation causes "necrosis" (death of tissue).[64]

The construction of these events in terms of a purpose that has failed is beautifully captured in a standard text for medical students (a text otherwise noteworthy for its extremely objective, factual descriptions) in which a discussion of the events covered in the last paragraph (sloughing, hemorrhaging) ends with the statement "When fertilization fails to occur, the endometrium is shed, and a new cycle starts. This is why it used to be taught that 'menstruation is the uterus crying for lack of a baby.'"[65]

I am arguing that just as seeing menopause as a kind of failure of the authority structure in the body contributes to our negative view of it, so does seeing menstruation as failed production contribute to our negative view of it. We have seen how Sontag describes our horror of production gone out of control. But another kind of horror for us is *lack* of production: the disused factory, the failed business, the idle machine. In his analysis of industrial civilization, Winner terms the stopping and breakdown of technological systems in modern society "apraxia" and describes it as "the ultimate horror, a condition to be avoided at all costs."[66] This horror of idle workers or machines seems to have been present even at earlier stages of industrialization. A nineteenth-century inventor, Thomas Ewbank, elaborated his view that the whole world "was designed for a Factory."[67] "It is only as a Factory, a *General Factory,* that the whole materials and influences of the earth are to be brought into play."[68] In this great workshop, humans' role is to produce: "God employs no idlers—creates none."[69]

> Like artificial motors, we are created for the work we can do—for the useful and productive ideas we can stamp upon matter. Engines running daily without doing any work resemble men who live without labor; both are spendthrifts dissipating means that would be productive if given to others.[70]

Menstruation not only carries with it the connotation of a productive system that has failed to produce, it also carries the idea of production gone awry, making products of no use, not to specification, unsalable, wasted, scrap. However disgusting it may be, menstrual blood will come out. Production gone awry is also an image that fills us with dismay and horror. Amid the glorification of machinery common in the nineteenth century were also fears of what machines could do if they went out of control. Capturing this fear, one satirist wrote of a steam-operated shaving machine that "sliced the noses off too many customers."[71] This image is close to the one Melville created in "The Bell-Tower," in which an inventor, who can be seen as an allegory of America, is killed by his mechanical slave,[72] as well as to Mumford's sorcerer's apprentice applied to modern machinery:[73]

> Our civilization has cleverly found a magic formula for setting both industrial and academic brooms and pails of water to work by themselves, in ever-increasing quantities at an ever-increasing speed. But we have lost the Master Magician's spell for altering the tempo of this process, or halting it when it ceases to serve human functions and purposes.[74]

Of course, how much one is gripped by the need to produce goods efficiently and properly depends on one's relationship to those goods. While packing pickles on an assembly line, I remember the foreman often holding up improperly packed bottles to us workers and trying to elicit shame at the bad job we were doing. But his job depended on efficient production, which meant many bottles filled right the first time. This factory did not yet have any effective method of quality control, and as soon as our supervisor was out of sight, our efforts went toward filling as few bottles as we could while still concealing who had filled which bottle. In other factories, workers seem to express a certain grim pleasure when they can register objections to company policy by enacting imagery of machinery out of control. Noble reports an incident in which workers resented a supervisor's order to "shut down their machines, pick up brooms, and get to work cleaning the area. But he forgot to tell them to stop. So, like the sorcerer's apprentice, diligently and obediently working to rule, they continued sweeping up all day long."[75]

Perhaps one reason the negative image of failed production is attached to menstruation is precisely that women are in some sinister sense out of control when they menstruate. They are not reproducing, not continuing the species, not preparing to stay at home with the baby, not providing a safe, warm womb to nurture a man's sperm. I think it is plain that the negative power behind the image of failure to produce can be considerable when applied metaphorically to women's bodies. Vern Bullough comments optimistically that "no reputable scientist today would regard menstruation as pathological,"[76] but this paragraph from a recent college text belies his hope:

> If fertilization and pregnancy do not occur, the corpus luteum degenerates and the levels of estrogens and progesterone decline. As the levels of these hormones decrease and their stimulatory effects are withdrawn, blood vessels of the endometrium undergo prolonged spasms (contractions) that reduce the bloodflow to the area of the endometrium supplied by the vessels. The resulting lack of blood causes the tissues of the affected region to degenerate. After some time, the vessels relax, which allows blood to flow through them again. However, capillaries in the area have become so weakened that blood leaks through them. This blood and the deteriorating endometrial tissue are discharged from the uterus as the menstrual flow. As a new ovarian cycle begins and the level of estrogens rises, the functional layer of the endometrium undergoes repair and once again begins to proliferate.[77]

In rapid succession the reader is confronted with "degenerate," "decline," "withdrawn," "spasms," "lack," "degenerate," "weakened," "leak," "deteriorate," "discharge," and, after all that, "repair."

In another standard text, we read:

> The sudden lack of these two hormones [estrogen and progesterone] causes the blood vessels of the endometrium to become spastic so that blood flow to the surface layers of the endometrium almost ceases. As a result, much of the endometrial tissue dies and sloughs into the uterine cavity. Then, small amounts of blood ooze from the denuded endometrial wall, causing a blood loss of about 50 ml during the next few days. The sloughed endometrial tissue plus the blood and much serous exudate from the denuded uterine surface, all together called the *menstrum,* is gradually expelled by intermittent contractions of the uterine muscle for about 3 to 5 days. This process is called *menstruation.*[78]

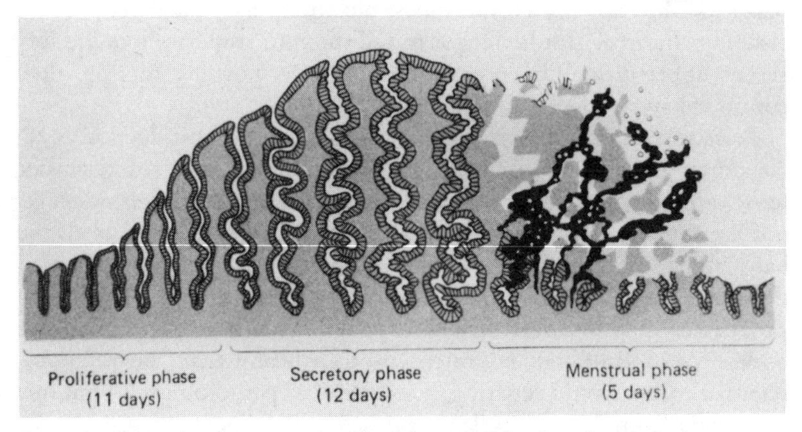

Fig. 9 An illustration from a current physiology text showing changes in the endometrium during the monthly cycle. The menstrual phase is represented visually as disintegration of the uterine lining. (Guyton 1984:624. Copyright © 1984 by CBS College Publishing. Reprinted by permission of CBS College Publishing.)

The illustration that accompanies this text (see Figure 9) captures very well the imagery of catastrophic disintegration: "ceasing," "dying," "losing," "denuding," and "expelling."

These are not neutral terms; rather, they convey failure and dissolution. Of course, not all texts contain such a plethora of negative terms in their descriptions of menstruation. But unacknowledged cultural attitudes can seep into scientific writing through evaluative words. Coming at this point from a slightly different angle, consider this extract from a text that describes male reproductive physiology. "The mechanisms which guide the *remarkable* cellular transformation from spermatid to mature sperm remain uncertain . . . Perhaps the most *amazing* characteristic of spermatogenesis is its *sheer magnitude:* the normal human male may manufacture several hundred million sperm per day (emphasis added)."[79] As we will see, this text has no parallel appreciation of female processes such as menstruation or ovulation, and it is surely no accident that this "remarkable" process involves precisely what menstruation does not in the medical view: production of something deemed valuable. Although this text sees such massive sperm production as unabashedly positive, in fact, only about one out of every 100 billion sperm ever makes it to fertilize an egg: from the very same point of view that sees menstruation as a waste product, surely here is something really worth crying about!

When this text turns to female reproduction, it describes menstruation in the same terms of failed production we saw earlier.

The fall in blood progesterone and estrogen, which results from *regression* of the corpus luteum, *deprives* the highly developed endometrial lining of its hormonal support; the immediate result is *profound constriction* of the uterine blood vessels due to production of vasoconstrictor prostaglandins, which leads to *diminished* supply of oxygen and nutrients. *Disintegration* starts, and the entire lining (except for a thin, deep layer which will regenerate the endometrium in the next cycle) begins to slough . . . The endometrial arterioles dilate, resulting in *hemorrhage* through the weakened capillary walls; the menstrual flow consists of this blood mixed with endometrial *debris* . . . The menstrual flow ceases as the endometrium *repairs* itself and then grows under the influence of rising blood estrogen concentration. [Emphasis added.][80]

And ovulation fares no better. In fact part of the reason ovulation does not merit the enthusiasm that spermatogenesis does may be that all the ovarian follicles containing ova are already present at birth. Far from being *produced* as sperm is, they seem to merely sit on the shelf, as it were, slowly degenerating and aging like overstocked inventory.

At birth, normal human ovaries contain an estimated one million follicles, and no new ones appear after birth. Thus, in marked contrast to the male, the newborn female already has all the germ cells she will ever have. Only a few, perhaps 400, are destined to reach full maturity during her active productive life. All the others degenerate at some point in their development so that few, if any, remain by the time she reaches menopause at approximately 50 years of age. One result of this is that the ova which are released (ovulated) near menopause are 30 to 35 years older than those ovulated just after puberty; it has been suggested that certain congenital defects, much commoner among children of older women, are the result of aging changes in the ovum.[81]

How different it would sound if texts like this one stressed the vast excess of follicles produced in a female fetus, compared to the number she will actually need. In addition, males are also born with a complement of germ cells (spermatogonia) that divide from time to time, and most of which will eventually differentiate into sperm. This text could easily discuss the fact that these male germ cells and their progeny are also subject to aging, much as female germ cells are. Although we would still be operating within the terms of the production metaphor, at least it would be applied in an evenhanded way to both males and females.

One response to my argument would be that menstruation just *is* in some objective sense a process of breakdown and deterioration. The particular words are chosen to describe it because they best fit the reality of what is happening. My counterargument is to look at other processes in the body that are fundamentally analogous to menstruation in that they involve the shedding of a lining to see whether they also are described in terms of breakdown and deterioration. The lining of the stomach, for example, is shed and replaced regularly, and seminal fluid picks up shedded cellular material as it goes through the various male ducts.

The lining of the stomach must protect itself against being digested by the hydrochloric acid produced in digestion. In the several texts quoted above, emphasis is on the *secretion* of mucus,[82] the *barrier* that mucous cells present to stomach acid,[83] and—in a phrase that gives the story away—the periodic *renewal* of the lining of the stomach.[84] There is no reference to degenerating, weakening, deteriorating, or repair, or even the more neutral shedding, sloughing, or replacement.

> The primary function of the gastric secretions is to begin the digestion of proteins. Unfortunately, though, the wall of the stomach is itself constructed mainly of smooth muscle which itself is mainly protein. Therefore, the surface of the stomach must be exceptionally well protected at all times against its own digestion. This function is performed mainly by mucus that is secreted in great abundance in all parts of the stomach. The entire surface of the stomach is covered by a layer of very small *mucous cells,* which themselves are composed almost entirely of mucus; this mucus prevents gastric secretions from ever touching the deeper layers of the stomach wall.[85]

In this account from an introductory physiology text, the emphasis is on production of mucus and protection of the stomach wall. It is not even mentioned, although it is analogous to menstruation, that the mucous cell layers must be continually sloughed off (and digested). Although all the general physiology texts I consulted describe menstruation as a process of disintegration needing repair, only specialized texts for medical students describe the stomach lining in the more neutral terms of "sloughing" and "renewal."[86] One can choose to look at what happens to the lining of stomachs and uteruses negatively as breakdown and decay needing repair or positively as continual production and replenishment. Of these two sides of the same coin, stomachs, which women *and* men have, fall on the positive side; uteruses, which only women have, fall on the negative.

One other analogous process is not handled negatively in the general physiology texts. Although it is well known to those researchers who work with male ejaculates that a very large proportion of the ejaculate is composed of shedded cellular material, the texts make no mention of a shedding process let alone processes of deterioration and repair in the male reproductive tract.[87]

What applies to menstruation once a month applies to menopause once in every lifetime. As we have seen, part of the current imagery attached to menopause is that of a breakdown of central control. Inextricably connected to this imagery is another aspect of failed production. Recall the metaphors of balanced intake and outgo that were applied to menopause up to the mid-nineteenth century, later to be replaced by metaphors of degeneration. In the early 1960s, new research on the role of estrogens in heart disease led to arguments that failure of female reproductive organs to produce much estrogen after menopause was debilitating to health.

This change is marked unmistakably in successive editions of a major gynecology text. In the 1940s and 1950s, menopause was described as usually not entailing "any very profound alteration in the woman's life current."[88] By the 1965 edition dramatic changes had occurred: "In the past few years there has been a radical change in viewpoint and some would regard the menopause as a possible pathological state rather than a physiological one and discuss therapeutic prevention rather than the amelioration of symptoms."[89]

In many current accounts, menopause is described as a state in which ovaries fail to produce estrogen.[90] The 1981 World Health Organization report defines menopause as an estrogen-deficiency disease.[91] Failure to produce estrogen is the leitmotif of another current text: "This period during which the cycles cease and the female sex hormones diminish rapidly to almost none at all is called the *menopause*. The cause of the menopause is the 'burning out' of the ovaries . . . Estrogens are produced in subcritical quantities for a short time after the menopause, but over a few years, as the final remaining primordial follicles become atretic, the production of estrogens by the ovaries falls almost to zero." Loss of ability to produce estrogen is seen as central to a woman's life: "At the time of the menopause a woman must readjust her life from one that has been physiologically stimulated by estrogen and progesterone production to one devoid of those hormones."[92]

Of course, I am not implying that the ovaries do not indeed produce much less estrogen than before. I am pointing to the choice of these textbook authors to emphasize above all else the negative aspects of ovaries failing to produce female hormones. By contrast, one current text shows us a positive view of the decline in estrogen production: "It would seem that although menopausal women do have an estrogen milieu which is lower than that necessary for *reproductive* function, it is not negligible or absent but is perhaps satisfactory for *maintenance* of *support tissues*. The menopause could then be regarded as a physiologic phenomenon which is protective in nature—protective from undesirable reproduction and the associated growth stimuli."[93]

I have presented the underlying metaphors contained in medical descriptions of menopause and menstruation to show that these ways of describing events are but one method of fitting an interpretation to the facts. Yet seeing that female organs are imagined to function within a hierarchical order whose members signal each other to produce various substances, all for the purpose of transporting eggs to a place where they can be fertilized and then grown, may not provide us with enough of a jolt to begin to see the contingent nature of these descriptions. Even seeing that the metaphors we choose fit very well with traditional roles assigned to women may still not be enough to make us question whether there might be another way to represent the same biological phenomena. In the following chapters I examine women's ordinary experience of menstruation and menopause looking for alternative visions.[94] And here I suggest some other ways that these physiological events could be described.

First, consider the teleological nature of the system, its assumed goal of implanting a fertilized egg. What if a woman has done everything in her power to avoid having an egg implant in her uterus, such as birth control or abstinence from heterosexual sex. Is it still appropriate to speak of the single purpose of her menstrual cycle as dedicated to implantation? From the woman's vantage point, it might capture the sense of events better to say the purpose of the cycle is the production of menstrual flow. Think for a moment how that might change the description in medical texts: "A drop in the formerly high levels of progesterone and estrogen creates the appropriate environment for reducing the excess layers of endometrial tissue. Constriction of capillary blood vessels causes a lower level of oxygen and nutrients and paves the way for a vigorous production of menstrual fluids. As a part of the renewal of the remaining endometrium, the capillaries begin to reopen, contributing some blood and serous fluid to the volume of endometrial material already beginning to flow." I can see no reason why the menstrual blood itself could not be seen as the desired "product" of the female cycle, except when the woman intends to become pregnant.

Would it be similarly possible to change the nature of the relationships assumed among the members of the organization—the hypothalamus, pituitary, ovaries, and so on? Why not, instead of an organization with a controller, a team playing a game? When a woman wants to get pregnant, it would be appropriate to describe her pituitary, ovaries, and so on as combining together, communicating with each other, to get the ball, so to speak, into the basket. The image of hierarchical control could give way to specialized function, the way a basketball team needs a center as well as a defense. When she did not want to become pregnant, the purpose of this activity could be considered the production of menstrual flow.

Eliminating the hierarchical organization and the idea of a single purpose to the menstrual cycle also greatly enlarges the ways we could think of menopause. A team which in its youth played vigorous soccer might, in advancing years, decide to enjoy a quieter "new game" where players still interact with each other in satisfying ways but where gentle interaction *itself* is the point of the game, not getting the ball into the basket—or the flow into the vagina.

4 Medical Metaphors of Women's Bodies: Birth

Man knows no more degrading or unbearable misery than forced labor.
—Friedrich Engels
The Condition of the Working Class in England

To understand the medical treatment of birth, we must recognize that in the development of western thought and medicine, the body came to be regarded as a machine. This mechanical metaphor got its start in seventeenth- and eighteenth-century French hospitals where the womb and uterus were spoken of "as though they formed a mechanical pump that in particular instances was more or less adequate to expel the fetus."[1] In the development of obstetrics, the metaphor of the uterus as a machine combines with the use of actual mechanical devices (such as forceps), which played a part in the replacement of female midwives' hands by male hands using tools.[2] (See Figures 10–13.) is often claimed that the metaphor of the body as a machine continues to dominate medical practice in the twentieth century and both underlies and accounts for our willingness to apply technology to birth and to intervene in the process. The woman's body is the machine and the doctor is the mechanic or technician who "fixes" it:

> During the 1940s, 1950s, and 1960s, birth was the processing of a machine by machines and skilled technicians.[3]

> Antenatal care becomes maintenance and malfunction-spotting work, and obstetrical intervention in delivery equals the repair of mechanical faults with mechanical skills. Concretely, as well as ideologically, women appear

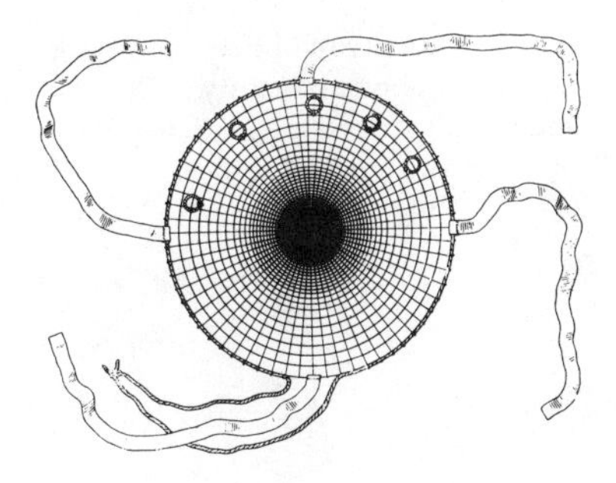

Fig. 10 Pierre Amand's sling for extracting the fetal head, introduced in the early eighteenth century. (Witkowski 1891, fig. 87, p. 139. Institute of the History of Medicine, the Johns Hopkins University.)

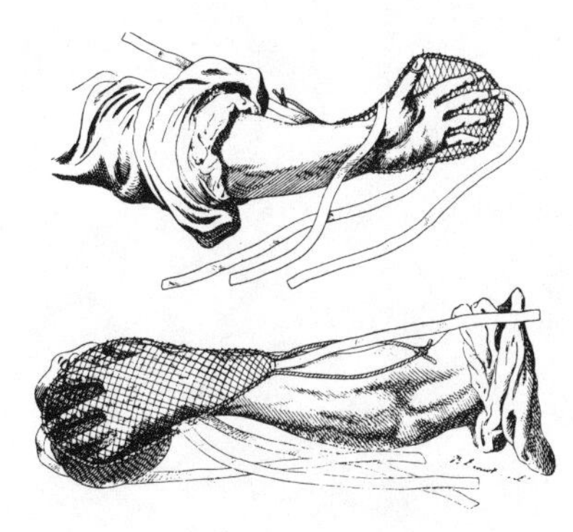

Fig. 11 The manner in which Amand's sling was applied. (Witkowski 1891, figs. 88, 89, p. 140. Institute of the History of Medicine, the Johns Hopkins University.)

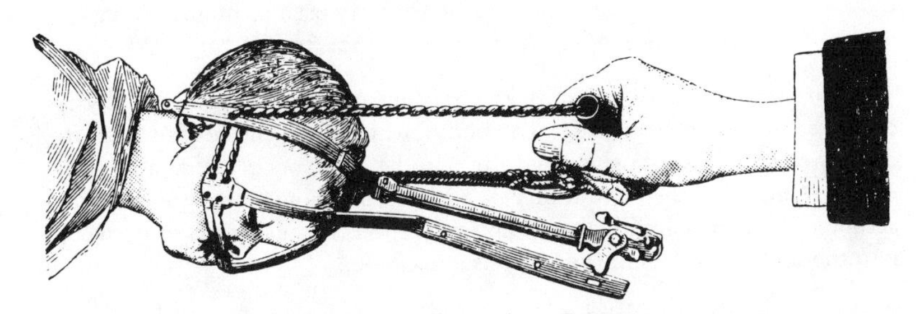

Fig. 12 The mid-nineteenth-century fetal extractor of Jules Poullet. (Witkowski 1891, fig. 129, p. 272. Institute of the History of Medicine, the Johns Hopkins University.)

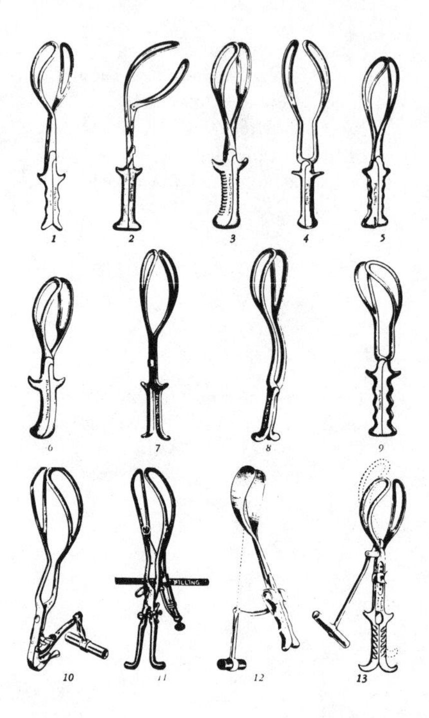

Fig. 13 A variety of obstetric forceps in use in the twentieth century. (Ullery and Castallo 1957, fig. 9–47. Reprinted by permission of F. A. Davis.)

to become machines, as machines are increasingly used to monitor pregnancy and labor and to initiate and terminate labor itself.[4]

> The Cartesian model of the body as a machine operates to make the physician a technician, or mechanic. The body breaks down and needs repair; it can be repaired in the hospital as a car is in the shop; once 'fixed,' a person can be returned to the community . . . Electronic monitoring was widely accepted in medicine with almost no reservations because it fits so perfectly into the medical model of the body as a machine.[5]

The metaphor of body as machine and doctor as mechanic may have been the dominant metaphor preoccupying medical practitioners and patients in the early development of scientific medicine. This is a matter for medical historians to investigate further, but at least the relatively low status of the early men who wielded forceps—"broken barbers, tailors, or even pork butchers" in the words of one disapproving midwife—fits comfortably with the notion that at that time the medical practitioner shared the social status of a mechanic.[6]

However, I think it is too simple to suggest that this metaphor alone continues to dominate medical practice. I would like to broaden the realm in which we look for metaphors in medicine to include the entire process of work in which the doctor is involved and the entire process of activity in which the woman is involved. In particular I would like to look carefully at whether analogies taken from the realm of production in factories are being applied to birth.[7]

Is reproduction being treated as a form of production, as we have seen menstruation and menopause treated as forms of failed production? The production metaphor, if it is involved, has quite different implications from the machine/mechanic metaphor. It allows us to ask, for example, whether the doctor is only a mechanic or perhaps more like a factory supervisor or even an owner. If the doctor is a supervisor, the woman might be a "laborer" whose "machine" (uterus) produces the "product," babies. To see what the elements of the metaphor are, it is necessary to look closely at the organization of factory production in our society to better understand whether processes occurring there might be said to occur in the realm of reproduction.[8] Broadening the scope of the inquiry about metaphors from the narrow—mechanic as doctor, machine as patient—also allows us to consider whether similar relationships of power and control operate in both realms of "production."

In a recent book about technology and labor, David Noble discusses how our society looks at technology as if it were an independent force that develops according to an autonomous process of evolution and to which we can do nothing but adjust:

> Our culture objectifies technology and sets it apart and above human affairs. Here technology has come to be viewed as an autonomous process, having a life of its own which proceeds automatically, and almost naturally, along a singular path. Supposedly self-defining and independent of social power and purpose, technology appears to be an external force impinging upon society, as it were, from outside, determining events to which people must forever adjust.[9]

Our focus on technology, its ever-changing needs and demands, diverts our attention from the social relationships of power and domination that are involved whenever humans use machines to produce goods in our society.

We can jolt ourselves out of our tendency to take technology as a given if we change the nature of its use. Noble suggests a fanciful example of an engineer who designed a new technical system that required, in order to function well, a great deal of control over the behavior of his fellow engineers in the laboratory. Such a design,

Noble suggests, might well be dismissed as ridiculous, whatever the merits of its components.

> But, if the same engineer created the same system for an industrial manager or the Air Force and required, for its successful functioning, control over the behavior of industrial workers or soldiers (or even engineers in their employ), the design might be deemed viable, even downright ingenious. The difference between the two situations is the power of the manager and the military to coerce workers and soldiers (and engineers) and the engineer's own lack of power to coerce his fellows.[10]

Imagining technology being used to control those who ordinarily use it to control others throws the power relationships into focus. Women involved in birth activist organizations, especially those who have experienced cesarean sections, see the power relationships in birth technology very clearly. Consider this example, which in some ways is parallel to Noble's engineer case:

> If your husband was told that he had to get an erection and ejaculate within a certain time or he'd be castrated, do you think it would be easy? To make it easier, perhaps he could have an I.V. put into his arm, be kept in one position, have straps placed around his penis, and be told not to move: He could be checked every few minutes; the sheet could be lifted to see if any "progress" had been made.[11]

This strikes us as ridiculous, but, as we will see, it may seem appropriate to us when women are placed in a structurally analogous position while giving birth.

My first task in this chapter is to show how the obstetrical literature looks at "labor." I mentioned briefly in Chapter 1 that since the uterus is seen as an involuntary muscle, it, rather than the woman, is seen as doing most of the labor. What the uterus does is expressed in terms that would be familiar to any student of time and motion studies used in industry to analyze and control workers' movements: "Labor is work; mechanically, work is the generation of motion against resistance. The forces involved in labor are those of the uterus and the abdomen that act to expel the fetus and that must overcome the resistance offered by the cervix to dilatation and the friction created by the birth canal during passage of the presenting part."[12] Note that the kind of "work" meant is mechanical work, as defined in physics, a narrow conception of force working against resistance. This is the same kind of breaking down of what is actually a complex and interrelated process that occurs when work such as screwing in a screw is broken down into simple mechanical processes by time and motion studies.

In the case of time and motion studies the object is clearly to control the exact movements of the worker so as to increase production.[13] The language applied to labor contractions sounds suspiciously as if doctors had the same object in mind. Uteruses produce "efficient or inefficient contractions";[14] good or poor labor is judged by the amount of "progress made in certain periods of time."[15] In the obstetrician Emanuel Friedman's well-known representations of average dilatation curves, the amount of time it takes a woman's cervix to open from 4 to 8 cm is described as a "good measure of the overall efficiency of the machine."[16] Presumably the "machine" referred to is the uterus.

Let us look more closely at how the uterus is held to a reasonable "progress," a certain "pace,"[17] and not allowed to stop and start with its natural rhythm.[18] A woman's labor, like factory labor, is subdivided into many stages and substages. The first stage includes the latent phase (slow effacement and dilatation of the cervix to 3 or 4 cm); the active phase (more rapid dilatation from 3 or 4 to 10 cm). The active phase is further subdivided into the phase of acceleration, the phase of maximum slope, and the phase of deceleration. The second stage includes the descent of the baby down the birth canal and its birth; and the third stage includes the separation and delivery of the placenta.[19]

Each stage and substage is assigned a rate of progression based on a statistical study of the rate characteristic of 95 percent of labors in the study. First stage, latent phase, should progress at 0.6 cm/hour; active phase, acceleration subphase, should progress at 0.6 cm/hour; active phase, subphase of maximum slope, should progress at 1.2 cm/hour or more for a first labor, 1.5 cm/hour for later labors; second stage should progress at 1 cm/hour and 2 cm/hour for a first and a second labor, respectively.[20] Deviation from these rates can produce a variety of "disorders": too slow a rate leads to "protracted" or "prolonged" latent phase, active phase, or deceleration phase; cessation of dilatation for specified periods leads to various kinds of "arrests."[21] Obstetrics handbooks list the proper management of these disorders: administration of medication (to sedate or stimulate); x-ray or sonogram to determine whether an obstruction is present; forceps or cesarean section.[22] Friedman's chart in one handbook shows graphically how many are the paths through arrest and protraction to cesarean section and how few the paths to vaginal delivery.[23] (See Figure 14.)

The language in which medical texts describe the effect of interventions often points directly to the idea that productivity is what is being increased. Amniotomy (breaking the amniotic sac) results in an increase in "work performed by the uterus"; oxytocin leads to a suc-

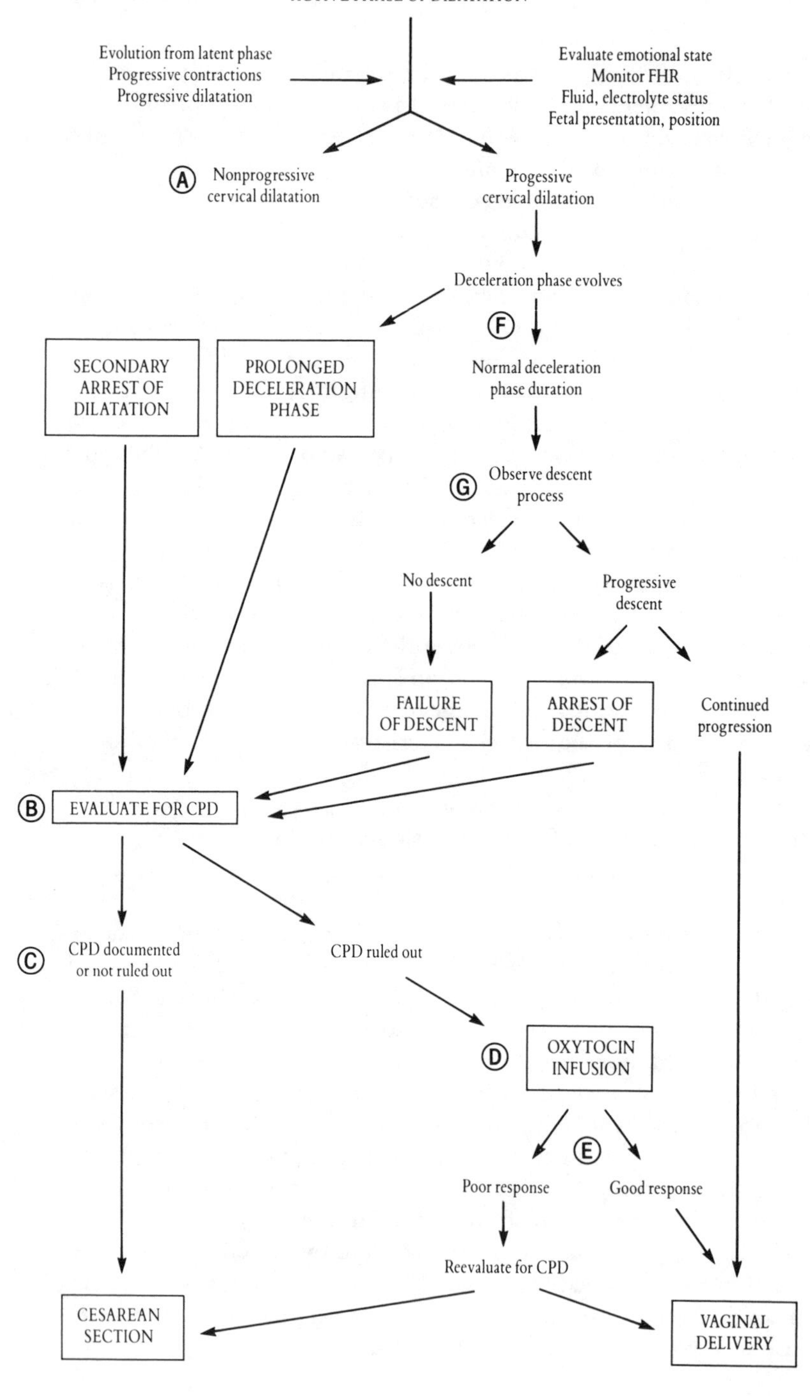
ACTIVE PHASE OF DILATATION
Evolution from latent phase
Progressive contractions
Progressive dilatation
Evaluate emotional state
Monitor FHR
Fluid, electrolyte status
Fetal presentation, position
A
Nonprogressive cervical dilatation
Progessive cervical dilatation
Deceleration phase evolves
F
SECONDARY ARREST OF DILATATION
PROLONGED DECELERATION PHASE
Normal deceleration phase duration
G
Observe descent process
No descent
Progressive descent
FAILURE OF DESCENT
ARREST OF DESCENT
Continued progression
B
EVALUATE FOR CPD
C
CPD documented or not ruled out
CPD ruled out
D
OXYTOCIN INFUSION
E
Poor response
Good response
Reevaluate for CPD
CESAREAN SECTION
VAGINAL DELIVERY

cessful completion of labor when "the uterus has simply become exhausted." Sometimes the uterus itself knows when to stop its effort; "it recognizes the futility of further effort (if the fetus is in an unusual posture or the pelvis is the wrong shape)."[24]

If the uterus is a machine being held to certain standards of efficient work, what is the woman's role? In the first stage of labor, she is seen as a passive host for the contracting uterus. An early (1950) edition of *Williams Obstetrics* says: "Uterine contractions are *involuntary.* Their action is not only independent of the mother's will, but of extra-uterine nervous control."[25] The latest edition (1985), a standard text used at Johns Hopkins Medical School, adds only that anesthesia sometimes interferes with contractions: "Uterine contractions are involuntary and, for the most part, independent of extrauterine control. Neural blockage from caudal or epidural anesthesia, if initiated quite early in labor, is sometimes associated with a reduction in the frequency and intensity of uterine contractions, but not after labor is well established."[26]

Despite assertions that uterine contractions are involuntary, at least some obstetrics texts contain direct evidence that this is not entirely so. As the 1950 *Williams Obstetrics* puts it, in a particularly contorted fashion,

> clinical observation indicates clearly that extraneous causes may interfere reflexly with the activity of the uterus. It is a matter of common experience that the entrance of the obstetrician into the lying-in chamber is frequently followed by a temporary cessation of the labor pains. Extreme nervousness, profound mental emotions, or excruciating pain may have a similar effect.[27]

This clinician's observation that labor contractions are affected by a woman's environment and emotional state has been omitted from recent editions of *Williams Obstetrics*.[28]

However, a similar statement can be found in another standard text used at Johns Hopkins: "Of utmost importance is that the physician take a reassuring, kindly attitude toward the patient. The degree of rapport that can be established with her will frequently determine the degree of comfort with which labor is experienced and perhaps also

---

Fig. 14 (*opposite*) A chart from a handbook for quick reference in hospitals that illustrates the great variety of ways a woman's labor can fail to progress adequately and so end up as a cesarean section. (E. A. Friedman, *Obstetrical Decision Making* [Toronto: B. C. Decker, 1982], 173.)

the length of labor."[29] And specific studies have shown that aspects of the environment such as light, noise, and movement from one place to another affect the length of labor in animals and humans.[30] Amazingly, this knowledge is not brought to bear on obstetrical treatment. For example, if a woman's labor slows down because her contractions are not sufficiently strong (hypotonic uterine dysfunction), most obstetrics texts suggest these causes: the pelvis is too small; the fetus is not positioned properly; or the uterus is too distended.[31] Nowhere is it suggested that the woman's general state of mind (fear, anxiety) might have led her to stop her labor, even though *"in many—perhaps one-half—of instances the cause of uterine dysfunction is unknown."*[32]

Seeing uterine contractions as "involuntary" has implications for the recommended treatment, of course.[33] The only measures obstetrics texts suggest are external manipulations of the woman's body: rupturing the amniotic sac, an oxytocin drip, which acts chemically on the uterus, or a cesarean section, which makes further contractions unnecessary.[34]

Given the view that uterine contractions are involuntary and that they are not accompanied by abdominal contractions in the first stage, the woman herself hardly has a role at this point. Contradictory as it may be, however, the woman is often evaluated during the first stage of labor as if she were the one performing—told she is doing well, or not well enough, told she can do it or cannot possibly do it, as well as warned about how much time remains in each part of the first stage and how well or poorly she is progressing. It is because the woman is really thought of as someone to control that scientific management strategies are thought to be appropriate. If there were only a machine lying there, one would simply turn it on and fix it if it broke, not subdivide its movements and exhort it to hold to a certain rate of production. In continuous-process industries where machines perform most of the work automatically, scientific management is no longer necessary, and control over the few workers left is greatly relaxed.[35]

Contradictions also appear in descriptions of the second stage of labor. Although this is when a woman is thought to "push" with her abdominal muscles along with the contracting uterus, sometimes it is hard to even detect a person present in the complex medical terminology. In the seventeenth edition of *Williams Obstetrics* (1985), the second stage of labor is described like this: "After the cervix is dilated fully, the force that is principally important in the expulsion of the fetus is that produced by increased intraabdominal pressure created by contraction of the abdominal muscles simultaneously with forced respiratory efforts with the glottis closed. This is usually referred to as 'pushing.'"[36] Despite the highly abstract nature of this description, it is apparent in the example of paraplegic women, that it is the *woman* who does the "pushing." In paraplegic women, who do not feel the urge to push, expulsion of the infant is said to be rarely possible unless the obstetrician detects the start of a contraction by feeling the woman's uterus and then instructing her to bear down with her abdominal muscles. In the second stage, a woman who can feel the urge to push is often exhorted to push harder and told she is approaching the end of the time allowed for this second stage. (Contrariwise, she may be told *not* to push.) Here she is quite clearly seen as the "laborer."[37]

One woman expressed the confusion she felt over the contradictory instructions she was given:

> And then there was that awful stage when they were telling me not to push and I couldn't [stop pushing]. You can't prevent yourself. One minute they're telling you that your uterus is an involuntary muscle and the next minute they're telling you not to push. I don't know whether you push with your uterus; I don't suppose you do.[38]

In sum, medical imagery juxtaposes two pictures: the uterus as a machine that produces the baby and the woman as laborer who produces the baby. Perhaps at times the two come together in a consistent form as the woman-laborer whose uterus-machine produces the baby.

What role is the doctor given? I think it is clear he is predominantly seen as the supervisor or foreman of the labor process. First, a nearly universal expression of the doctor's role is that he "manages" labor. Special conditions require "active management";[39] previous cesarean section or other complications call forth "expectant management"[40] if a vaginal birth is being allowed. Second, it is of course the doctors who decide when the "pace" of work is insufficient and warrants speeding up by drugs or mechanical devices. Kieran O'Driscoll, an Irish obstetrician who is a strong advocate of "active management," has instituted a rigid scheme to control labor in the National Maternity Hospital in Dublin. As Ann Oakley summarizes,

> The essence of active management is that no labour is allowed to last beyond 12 hours. (The graphs used for recording labour and delivery data in the National Maternity Hospital do not allow for recordings over more than a 12-hour period.) In primigravidae [first-time mothers], active management of labour runs as follows: ARM (artificial rupture of membranes) is performed 1 hour after the initial diagnosis of labour "unless dilatation of cervix

is proceeding at a satisfactory rate." One hour after this unless the dilatation of the cervix has speeded up, oxytocin begins to be given intravenously: "There is a standard procedure, applied in all circumstances and by every member of staff, as follows: 10 units of oxytocin in 1 litre of 5 per cent dextrose solution is used."[41]

If the doctor is managing the uterus as machine and the woman as laborer, is the baby seen as a "product"? It seems beyond reproach for doctors to be concerned with the "fetal outcome" of a birth. What seems significant is that cesarean section, which requires the most "management" by the doctor and the least "labor" by the uterus and the woman, is seen as providing the best products.[42] "Doctors have created the attitude that a cesarean delivery implies a perfect baby."[43] Some accounts even celebrate the "dramatic increase" in cesarean rates. "Credit" should be given to earlier colleagues who advocated the increased rate (from 3 to 8 percent ten years ago to 15 to 23 percent today) and fostered a change in attitude (clearly one to be desired): "formerly focus was on the mother and delivery, and now it is centered on fetal outcome."[44] It follows that the current rate may not be as high as it will go. The *New York Times* quotes Dr. Robert Sokol: "'Doctors are not going to hesitate to do a Caesarean if there is any possibility it could improve the outcome.' By this criterion alone, he said, 'a 20 percent rate might not be too high.'"[45] Others detect a "new principle": "The long-held concept of vaginal delivery is rapidly giving way. The new growing principle seems to be, *vaginal delivery only of selected patients*."[46]

This belief that cesarean sections produce better-quality babies may be partly related to the attitude that even normal labor is intrinsically traumatic to the baby. From the early description by a nineteenth-century gynecologist of the uterus as a death missile,[47] through later (1920) descriptions of labor as being like the mother falling on a pitchfork or the baby's head being caught in a door jamb,[48] to contemporary efforts of obstetricians to ease the terrible experience of birth for the infant by dim lights and warm baths after birth, a role is constructed for the doctor to ally with the baby against the potential destruction wreaked on it by the mother's body. In Rothman's terms, "mother/fetus are seen in the medical model as a conflicting dyad rather than as an integral unit."[49]

In light of these images, extracting the baby by surgery could easily be seen as the baby's only saving grace. In fact, women are consoled after a cesarean which has made them angry and disappointed that they should be happy they have a healthy baby. Focusing on the product of the labor, of course, ignores what the woman may have been equally concerned about: the nature of her own experience of the birth. One woman reacted with redoubled anger at the obliteration of her own experience. Her experience of the cesarean was one of loss and disappointment at missing the birth she had planned for: "missing seeing her weighed, her first bath, little things. It was scary lying there having been cut open and you can't move." In addition, the recovery period was frustrating and uncomfortable: "I couldn't take care of the baby, they wouldn't let my husband stay overnight, they wouldn't let me eat. Going hours without food, I'm supposed to heal from major surgery, who could heal when they've got no nutrients? When they brought the soft tray I could have thrown it across the room." In this context, she became infuriated when people (nurses, doctors, and family) told her she was lucky to have a healthy baby. In a childbirth education class she attended before the birth of her next child, she became furious at a couple who mentioned that their first priority was to have a healthy baby. "I jumped all over them. I said that goes without saying: you shouldn't even have to mention that. Everybody wants a healthy baby. That's just not worth mentioning, you wouldn't get pregnant in the first place if you didn't want a healthy baby. [And mockingly] 'Oh, gee, we were really hoping for this child to be handicapped! I was so disappointed!' You know, think about it" (Sarah Lasch).

This woman's anger came from feeling that her own experience of the birth counted for nothing next to the welfare of her baby. But the baby was never in distress and in fact emerged with an Apgar score of 9/10 (a perfect score is 10/10), so it was hard for her to give credence to the staff's concern for the baby's well-being. Another woman, who had had a previous cesarean section, was amazed at how little the resident who attended her at the hospital knew about women's feelings about cesareans. "He asked a lot of questions: 'Why didn't you want to have another cesarean? They cut you below, they cut you above, what's the difference?' It really showed me how naive they are: put this woman out of her misery and she'll be eternally grateful for us giving her her baby. I told him about my disappointment, missing the birth, and the pain afterward, and he said 'Really! Wow!'" (Laura Cromwell).

In our way of thinking, as we have seen, the sphere of home and the sphere of work are sharply divided. Labor at a factory seems very different to us than housework or a woman's labor in childbirth because the one is sought and paid for in the marketplace whereas the others are not. Perhaps for this reason, most contemporary accounts of production in preindustrial societies (where this division is not present, at least not in the same way) fail to develop a model of productive activity that is general enough to include the labor that women do inside the house, much less the labor that they do to produce children.[50] As Raymond Williams puts it:

> It is very remarkable that if you look across the whole gamut of Marxism, the material-physical importance of the human reproductive process has been generally overlooked. Correct and necessary points have been made about the exploitation of women or the role of the family, but no major account of this whole area is available. Yet it is scarcely possible to doubt the absolute centrality of human reproduction and nurture and the unquestioned physicality of it.[51]

It is even more remarkable, given that since the fifteenth century the same English word, from the same root, "labor," has been used to describe what women do in bringing forth children and what men and women do in producing things for use and exchange in the home and market.[52] All this contains a double irony. When anthropologists, armed with concepts of production and labor that they associate with work done primarily by men in factories and enterprises outside the home, try to describe preindustrial societies, they often completely overlook the labor that women do; when medical doctors describe the labor that women do in childbirth, their expectations center on how labor of other kinds is organized in our society and how technology and machinery can be used to control those who labor. In both cases women lose, in the first by being overlooked and in the second by having a complex process that interrelates physical, emotional, and mental experience treated as if it could be broken down and managed like other forms of production.

In Chapters 3 and 4 we have seen that the dominant medical metaphors applied to women's bodies in menstruation, birth, and menopause involve a hierarchical system of centralized control organized for the purpose of efficient production and speed. Medical attention usually is given when this system undergoes breakdown, decay, failure, or inefficiency. How might we respond to this situation?

Marxists have found the application of market principles to noncommodities disturbing. Marx himself saw this process as inevitable but regrettable. He spoke of things "that are in themselves no commodities, such as conscience, honour, &c., . . . being offered for sale by their holders, and of thus acquiring, through their price, the form of commodities. Hence, an object may have a price without having value. The price in that case is imaginary, like certain quantities in mathematics."[53] His term "imaginary price" captures his sense that these are not the sort of things a price can be put on. Similarly, Winner uses the term "perverse" for the process by which norms from the realm of production are inappropriately extended into other realms, "instances in which things have become senselessly or inappropriately efficient, speedy, rationalized, measured, or technically refined . . . The predominance of instrumental norms can be seen as a spillover or exaggeration of the development of technical means. It is not that such norms are perverse in themselves but rather that they have escaped their accustomed sphere."[54]

Feminists have also objected to production metaphors being used to describe reproduction. In *Woman's Estate,* Juliet Mitchell refers to reproduction as a "sad mimicry of production," thinking how parenthood can be seen as an imitation of work: "the child is seen as an object created by the mother, in the same way as a commodity is created by a worker."[55] Mitchell wrote this in 1971. By now, in the mid-1980s I would say, and she might agree, that we have not so much a sad mimicry of production as a destructive travesty.

These reactions from theoreticians lead us squarely into the central question of this book: How do women as they lead ordinary lives respond to scientific metaphors about their bodies? Do they perceive them in the first place? Do they accept them as natural and as women's rightful due or resign themselves to tolerating them? Or do they fight them as deep and sinister threats to a full existence?

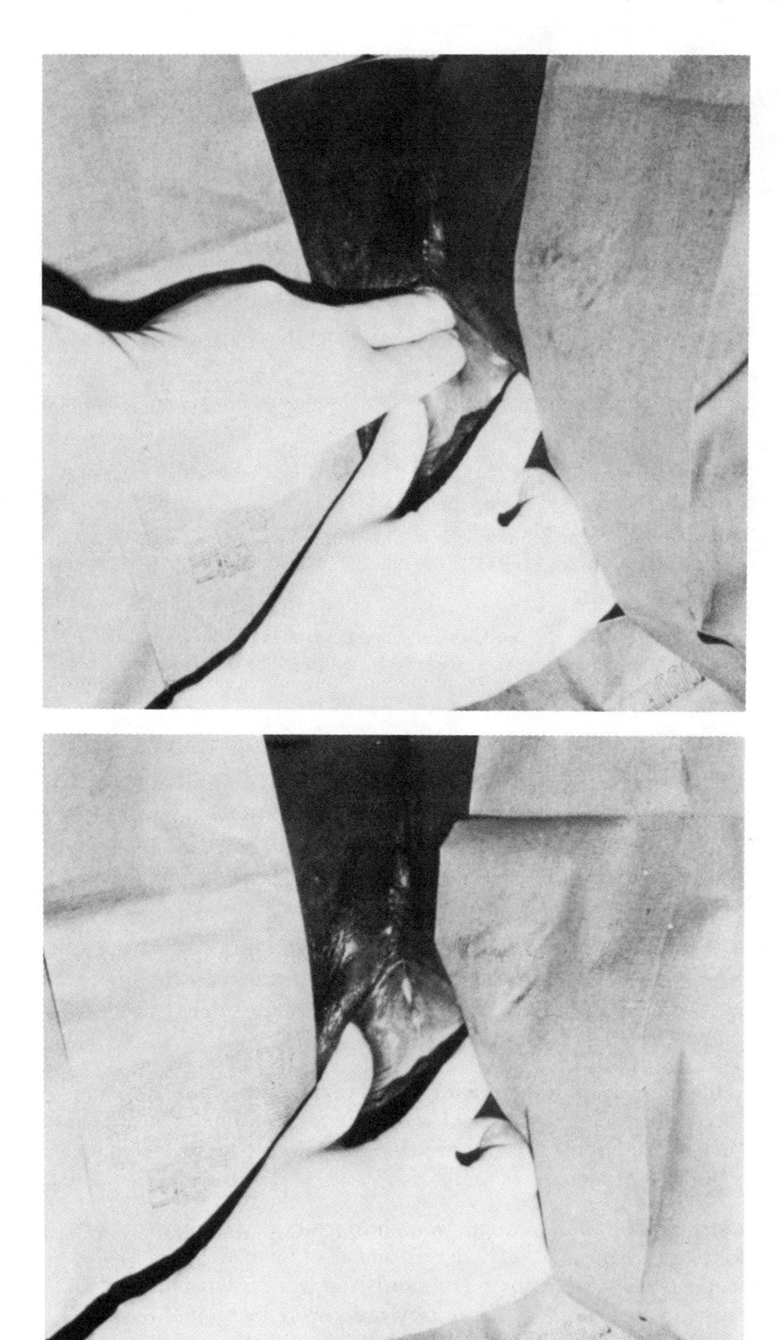

Fig. 15 The vaginal examination as performed during a woman's labor. (Pritchard and MacDonald 1980:412, figs. 17-1A and 17-1B. Reprinted by permission of Appleton-Century-Crofts.)

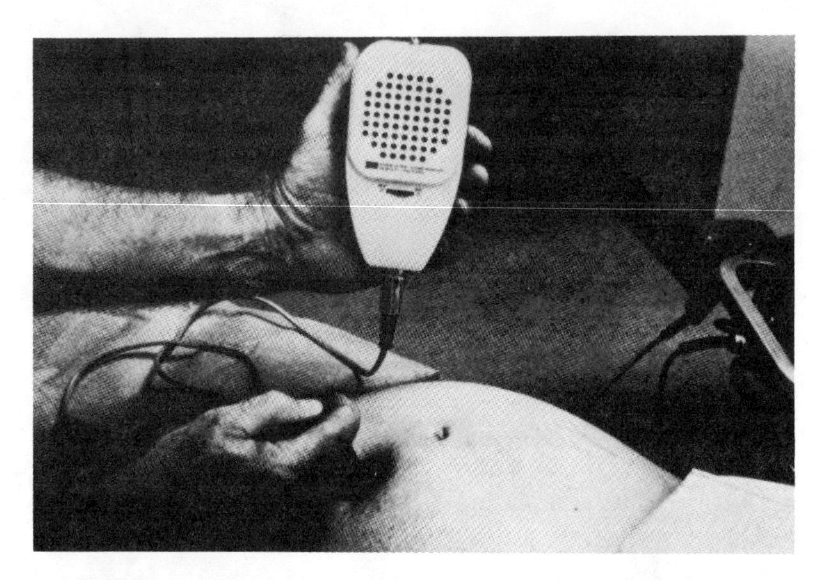

Fig. 16 The doptone, which uses ultrasound to monitor the fetal heart rate. The rubber strap shown across the mother's abdomen holds the transducer in place. (Pritchard and MacDonald 1980:412, fig. 17-2B. Reprinted by permission of Appleton-Century-Crofts.)

climax." She says it's almost the same thing: you're getting all ready to have this baby and then they switch you to this table and rush you down to the delivery room and say, "Okay, now you can do it." It's so intrusive and interrupts the whole thing; it's hard to get back to where you were. (Rebecca Lentz)

Many hospital maternity staffs administer an epidural, a form of spinal anesthesia, that numbs the woman from the waist down, makes her unable to push effectively, and usually necessitates delivery with forceps.

I asked for an epidural, not knowing that I was actually in transition and nearly fully dilated. At six o'clock I was ready to push, but with the epidural I couldn't feel the urge; we had to watch the monitor to know when to push. (Birth report)[2]

Oftentimes when you're having a baby the doctors act like they're watching TV. Your legs are up and you're draped and all of them are going like this [gestures in the air with a poking motion]. You wonder what they have down there, a portable T.V., or are they really working on me? *What* is going on? Half of you is numb so you can't feel what they're doing. And everybody just [mumbles] and you want to know, my God, what is going on? Then they start to talk in terms that you don't know and you think, I've got cancer, I'm bleeding to death, some *drastic* something. Whew! It's just too much sometimes; you need those few days of rest after the labor just to get your mind together. (Eileen Miller)

Before we can move from these discursive statements to women's underlying notions of self and body, we must have a clear idea of what it means to have a body image. Paul Schilder, an early Gestalt psychologist, gives us an idea of what happens when his patients have disturbed body images, in particular when these images lack wholeness. He argues that disturbed body images have some part in the development of almost every neurosis.[3] In one detailed example, "a case of loss of unity in the body-image," the patient reports:

> When I get this anxiety state I cannot walk further. I run into myself. It breaks me into pieces. I am like a spray. I lose my centre of gravity. I have no weight. I am quite mechanical. I have gone to pieces. I am like a marionette. I lack something to hold me together.[4]

Schilder describes other cases:

> The patient sees his face in the mirror changed, rigid, and distorted. His own voice seems strange and unfamiliar to him, and he shudders at the sound of it as if it were not himself speaking. Gottfried feels that his movements are interrupted. His body feels as if it were dead and he has the sensation that a dynamo is hissing in his head. The body feels too light, just as if it could fly. A patient of Loewy's says, "I feel the body not for me but for itself." The patients look for their limbs in the bed. A patient of Pick's complains that his eyes are like two holes through which he looks.[5]

Although these compelling experiences of bodily fragmentation frequently accompany neurosis, Schilder contends that they are never far from the everyday life of a normal person:

> The important conclusion we may draw is that feeling our body intact is not a matter of course. It is the effect of self-love. When destructive tendencies go on, the body is spread over the world . . . One would like to say that we lose the unity of our body only under special pathological conditions; but we also have to remember how much the feeling of our body varies under normal conditions. When we touch an object with a stick we feel with the end of the stick. We feel that clothes eventually become a part of ourselves. *We build the picture of our body again and again.* [Emphasis added.][6]

Schilder believes that loss of unity in the body image comes from the individual's experience in the family during the course of psychic development: "There are forces of hatred scattering the picture of our own body and forces of love putting it together."[7] What women say about their bodies forces us to look beyond the family to features of the social and cultural organization of experience that can also affect body image and can lead to fragmentations similar to those Schilder describes.

Since these forms of extreme fragmentation surely did not help Schilder's patients function in any optimal way, it was natural for him and other early Gestalt therapists to seek wholeness for these people. The psychiatrist Clifford Scott called the whole that was for the moment split, disorganized, or disintegrated "the body schema."[8] In extreme cases such as those of Schilder's patients, the shape of the whole might not be clear, but the shape of the parts that have come asunder is altogether clear. When people experience life in a more ordinary way, the problem is the opposite: how, under an appearance (and perhaps reality) of perfectly adequate functioning in the world, do we find out whether the self is whole or separated? More subtle techniques are needed, and the ones I have found most useful are linguistic.

Examining our everyday language, as George Lakoff and Mark Johnson's *Metaphors We Live By* has shown, allows us to identify certain metaphors that structure the way we think, talk, and act. For example, the conceptual metaphor "argument is war" is reflected in ordinary language in many ways:

Your claims are *indefensible*.

He *attacked every weak point* in my argument.

His criticisms were *right on target*.

I *demolished* his argument.[9]

So is the metaphorical concept "time is money":

You're *wasting* my time.

This gadget will *save* you hours.

I've *invested* a lot of time in her.

You need to *budget* your time.

He's living on *borrowed* time.

You don't use your time *profitably*.[10]

Applying this technique to our interviews, the central image women use is the following:

Your self is separate from your body.

This central image has a number of corollaries:

Your body is something your self has to adjust to or cope with.

One young woman described her early months of menstruation as "adjusting to how I have to cope with my body" (Kristin Lassiter). Women who experience serious menstrual discomfort feel at odds with the body they have to "cope" with: "I felt betrayed by my body when I started getting really serious or bad pains, I was irritated by my body, I felt betrayed and angry" (Ellie Yamada). Women who do not experience much discomfort still speak of their bodies as separate from their selves: "It doesn't really affect my body so I don't let it affect me mentally" (Mara Lenhart).

Your body needs to be controlled by your self.

Here separation is expressed as concern about the amount of control one has over one's body: "I resent the fact that menstruation changes so much about my body, my moods. I don't have enough control over that yet" (Abbie Phillips). Or, speaking about menopause, "I guess it's more of a fear, not of postmenopause, just of the actual process. You're not really in control of your body. It's not predictable, that's what scares me about it" (Tania Parrish).

Your body sends you signals.

This imagery was common in descriptions of labor. "So I just think it's amazing how your body tells you these things. I just started getting really tired. It's like my body was saying, 'All right you've had enough, now you stop.' It's really amazing. Your body tells you these things. Like, with not being able to sleep, I think, well maybe the baby's going to be getting up at this time anyway during the night, to get fed or whatever. So in a way it's preparing me for what's coming. Just like when you needed more sleep in the beginning. It's when the baby's developing the most anyway. So it sends out signals. Sometimes you pick them up, sometimes you don't" (Jan Moore).

Menstruation, menopause, labor, birthing and their component stages are states you go through or things that happen to you (not actions you do).

Menstruation is almost always described as "a process a woman goes through" (Kristin Lassiter) or something that happens: "I don't even think about it. It's just something that happens. It's not really good or bad to me anymore. When it first started happening, I was cursing because I had to put up with this. It's just sort of a process in my life" (Tania Parrish). Menopause is spoken of the same way: "I ignored it with such determination that I don't even know how old I was or how long it took me to get through it" (Martina Ostrov).

And the stages of labor and delivery are described in much the same terms. "Number one, I didn't think I could go through the process of delivery—that has always terrified me . . . I just can't see myself going through labor and delivery without some kind of anesthesia" (Chris Gonzales). "I mean you gotta go through the labor before you go through the delivery. I mean you've gone through all that, why not go through another hour or two?" (Jenny Powers). "Women have been going through this for millions of years. Even if it is terribly painful for five hours or whatever, I can go through it like everybody else did. Look at all the babies around" (Marcia Levy).

Menstruation, menopause and birth contractions are separate from the self. They are "*the* contractions," "*the* hot flashes" (not *mine*); they "come on"; women "get them."

Menstrual periods are spoken of as something a woman "gets" or "has"; in black English, they are something a woman is "on," they are even something she can "lose," consider "giving away" or "trading in." "I remember I had it and then I didn't have it, I skipped a month and that concerned me. I wouldn't give mine away because it's important to me if I ever want to have a family, which I probably will" (Kristin Lassiter).

Machine imagery often appears, as women wish "there was a way you could shut off your menstrual cycle like a faucet and then turn it back on again" (Rachel Lehman). "I wish I had a button right here that I could turn off and turn on when I wanted to" (Cathy Roark). "There should be some switch that you can psychically turn" (Meg O'Hara).

Similarly, in the case of labor and delivery: "In the beginning I didn't even know if it was true or false labor because there were continuous contractions. The contractions came from the back to the front like a girdle, but they were not regular like everyone else's. When a contraction would come on I would do my breathing" (Kathy Madsen). "While we were relaxing eating our dinner, the contractions were five minutes apart but very mild. While in the store the contractions were still strong and five to six minutes apart. Every time I would get a contraction Tom would massage my lower back (Birth report). "At 4:00 A.M. I had my first contraction. About seven minutes later I had another. Then they began to get stronger and came about every three minutes" (Birth report).

And in the case of menopause: "The only effect that I was aware of was that I got the flashes" (Martina Ostrov). "When I was fifty years old, whoosh, it was here and then nothing. It just stopped and that's about it. So funny to me, it wasn't here for one month, it wasn't here for two months, it wasn't here for three months, and I'm not pregnant, so what is it? So that's it, that's the end of the line" (Freda Von Hausen).

The separation of self from body that women describe when they talk about menstruation, menopause, and birth is present to an extreme degree when they describe cesarean section. Before this century, cesarean sections were performed exceedingly rarely, only in cases of direst emergency. (See Figures 17 and 18.) Today they continue in many cases to be life-saving emergency procedures, sparing women and infants death, suffering, and damage. (See Figures 19 A and B.) But questions have arisen over whether the current overall cesarean rate can be entirely justified by life-threatening situations. Official figures for 1983 place the rate of cesarean section in the United States at just over 20 percent of all births: one in five babies in this country are brought out of their mothers through an incision in the mother's abdomen. This represents a rise of 2 percent over the rate in 1982 (18.5 percent), and a rise of 7 percent over the much lower rate (13 percent) in 1977.[11] Some reasons for this steady and rapid rise in the rate of cesarean sections are simple to understand. For example, obstetricians and obstetrics textbooks justifiably stress the increased safety of the operation for both mother and baby; mortality and morbidity caused by the operation have decreased.[12] However, many women in health activist organizations and many obstetricians do not believe the current rate is justified; on the contrary, they argue that it is overall a detriment to the health and happiness of mothers, babies, and their families.

The lines of conflict are many. Whereas defenders of the increased rate link it to the decreasing perinatal mortality rate,[13] critics argue that there is no proof that the two factors are causally linked, observing that they are often not even correlated.[14] Further, institutions in

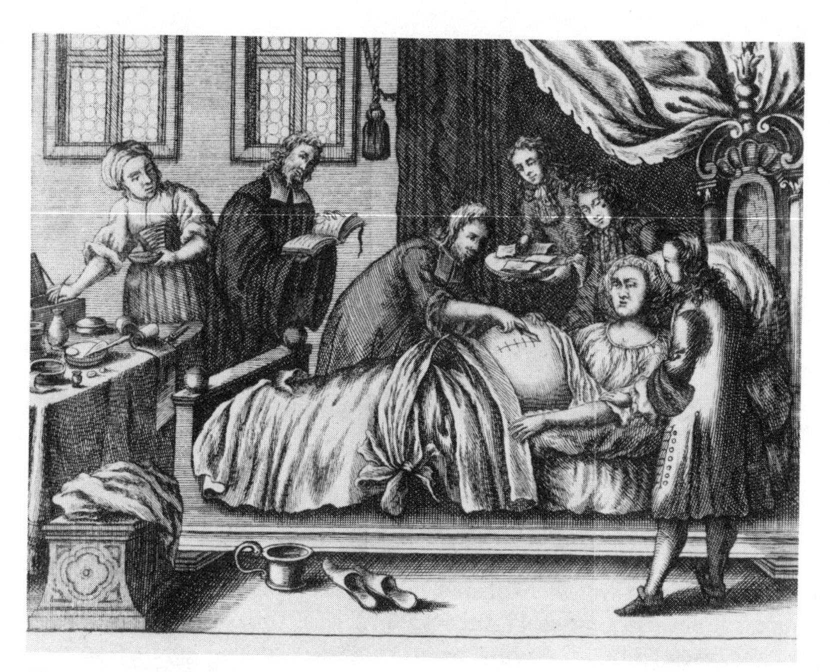

Fig. 17 A cesarean section in 1722. This operation was very rarely performed on the living in the seventeenth and eighteenth centuries. (Völter 1687. Institute of the History of Medicine, the Johns Hopkins University.)

other countries with a very low cesarean rate also have attained comparably low perinatal mortality rates, presumably because of better prenatal care.[15] Whereas defenders stress the increased safety of the procedure now as compared with earlier,[16] critics stress the comparative danger to both mother and baby of doing the procedure at all as compared with vaginal birth.[17] Whereas defenders stress the danger to the baby of unusual presentations (especially breech) and unusually long labors, critics stress the possibility of using simpler, nonoperative techniques to turn the baby or speed up the labor.[18] Whereas defenders stress the comfort of the mother (she can often completely avoid the experience of labor) and the baby (who does not have to be subjected to the process of squeezing through the birth canal), critics stress the sometimes extreme discomfort mothers suffer after the operation and the mother's relative inability to care for the infant while recovering from major surgery. Whereas defenders say they are forced to perform cesarean sections if there is any possibility of trouble in order to avoid

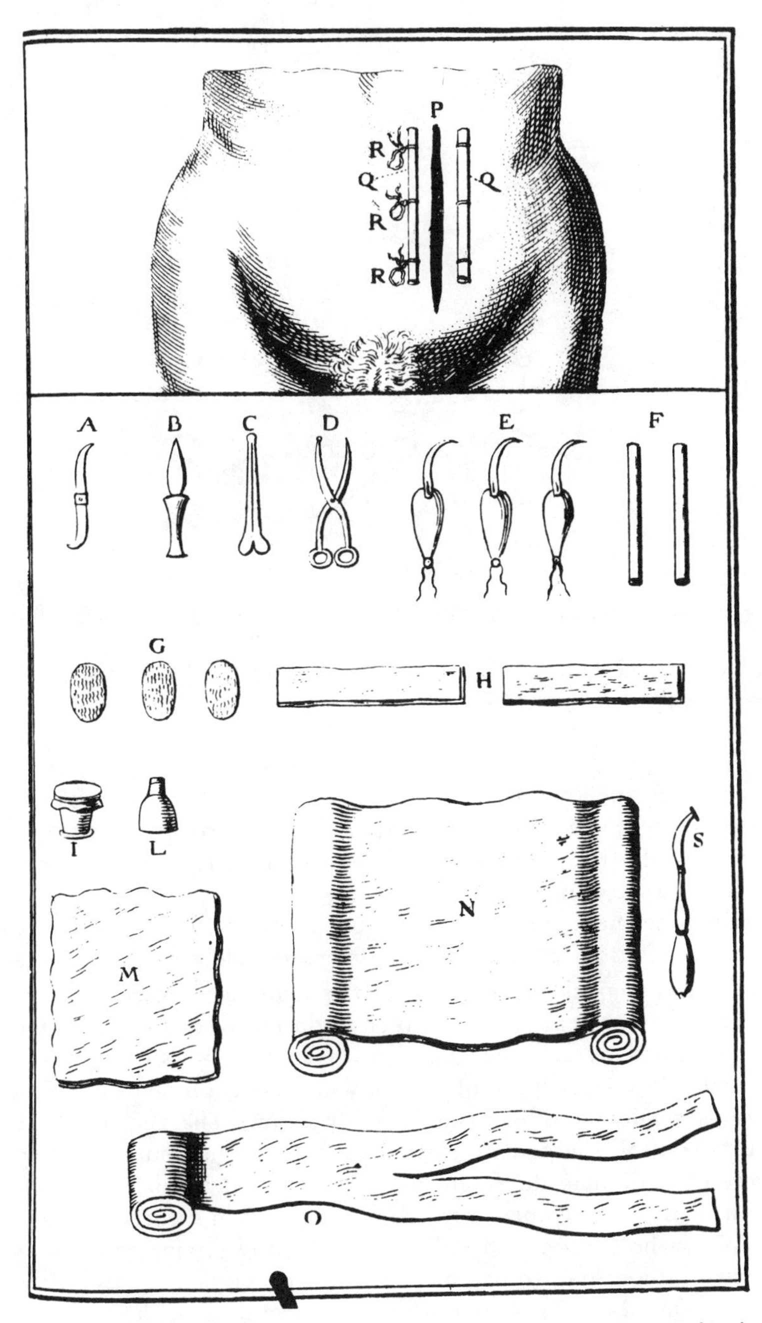

Fig. 18 Instruments, incision, sutures, and bandages for cesarean section used in the eighteenth century. (Mesnard 1743. Institute of the History of Medicine, the Johns Hopkins University.)

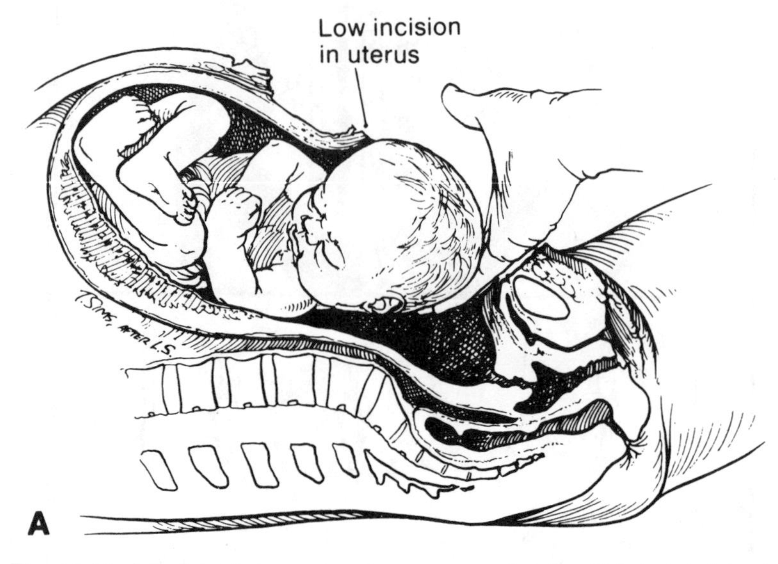

Fig. 19A Modern obstetrical procedure for cesarean section, in which the baby's head is lifted out through uterine and abdominal incisions.

malpractice suits (in this case more intervention rather than less intervention protects the doctor),[19] critics say informed consent can be used to protect the doctor and allow the woman more latitude in the kind of delivery she wants.[20]

Whatever the merits of each side of this debate, it is clear that cesarean sections affect women very differently and often more negatively than vaginal births.[21] One of the commonest reactions, especially when a woman had planned on vaginal delivery and the surgery occurred under conditions defined as an emergency, is to feel out of control: "I guess I felt out of control with a c-section, like everything was just totally out of my control" (Joan Tyson). The feeling of being out of control is an intense experience of fragmentation, akin to but more extreme than the fragmentation women experience in vaginal birth. Part of this comes from the necessity, given surgery, for more people to be touching, handling, cutting, and sewing one's body, some of whom the woman may never have seen before. "Somehow being referred to as a 'section' after a cesarean does not help you feel like a whole person. I felt even more like a fragmented body the next day when I thanked the doctor who helped in my cesarean for his

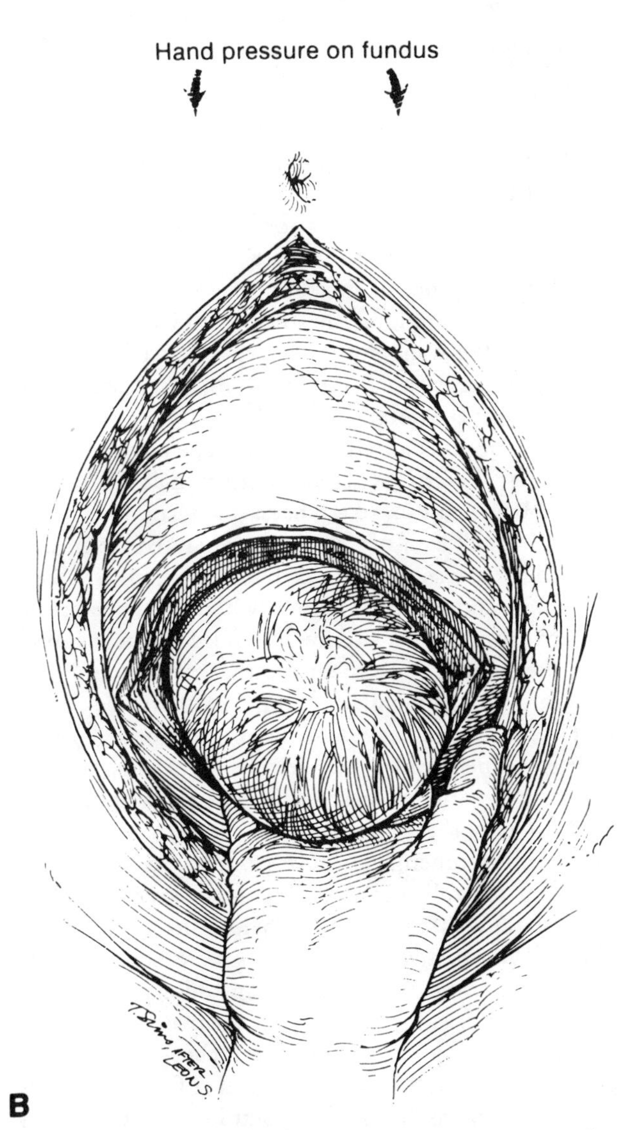

Fig. 19B As the baby's head is lifted, hand pressure is applied to the uterus through the abdominal wall to help push out the baby. (Pritchard and MacDonald 1980:1089, figs. 43-8A and 43-8B. Reprinted by permission of Appleton-Century-Crofts.)

assistance and he did not even remember who I was."[22] "They talked over me and to each other, but not to me. I felt like an object and not a human."[23] For such a woman, the separation between body and self needs no further explanation: she is literally divided into her "self" and her "body"; people are doing things to her body but paying no attention to her self. "I thought I would be a slab of meat someone would cut up, a far cry from my vision of an 'awake and aware' childbirth."[24] "When you have had a previous c-section, the first thing they have to do is rip the old scar out. They took this thing and threw it toward a bucket; it fell on the floor and someone asked if anyone wanted to go fishing!" (Pamela Hunter).

In part, this experience of splitting between self and body is simply a result of spinal or epidural anesthesia, which produces numbness from the neck or the waist down. As we have seen, women who give birth under epidural or spinal anesthesia also experience their selves as becoming an object the doctors manipulate. But in the case of cesarean section, the numbness is intensified by a drape placed across the woman's chest so that she cannot see her bottom half. "It was weird with the drape and being numb, like what a magician does when he cuts a lady in half."[25] "I felt as if my head was separated from my body . . . I couldn't believe it was me—it felt like a movie that I was watching . . . I felt strange and detached from my body."[26] "I wanted to scream. I felt detached from my own body . . . as if I was watching the whole thing from the ceiling."[27] Even an anesthetist who described her own experience in childbirth with nothing but gratitude for the "warm waves of pleasant paresthesia . . . lapping up both legs" comments that "my legs seemed completely detached from my body."[28]

Women may experience these feelings of fragmentation and objectification with either vaginal birth or cesarean section, but to my knowledge, only women who have cesarean section go farther and describe feeling forcibly violated. The images vary in the following examples, but they share a sense of helplessness while suffering terrible violence. *Crucifixion:* "I felt as if I were being crucified—both arms stretched out and pinned down spread-eagle."[29] *Rape:* "It was like a feeling of rape, completely out of control. My body was totally violated. I felt nobody cared. You wear cheap clothes and a lot of makeup and if you're raped, it's your fault. I felt the same thing with the cesarean" (Sarah Lasch). Women "shudder at the memory of being tied on the table and completely vulnerable to a man wielding a knife."[30] "I felt assaulted, raped perhaps, by what had happened to my body with absolutely no control on my part."[31] *Evisceration:* "The c-section was like the scene in the movie *Quest for Fire* where they showed a man being eviscerated alive and then having his guts eaten by other men" (Laura Cromwell).

These women are putting into words a feeling of alienation between themselves and the event of cesarean section, akin to the alienation of the laborer from his or her work, but beside which the separation of laborer and work pales. What about the "product" of these labors? It is frequently recognized in the women's health movement literature that when a woman's production of a baby is interfered with, especially when she does not birth the child herself but has it extracted surgically by a doctor, she feels detached from the child after its birth.[32] Even before a planned cesarean women express separation from both the event and the baby in the common phrase "They are going to take the baby next week."[33] The ways in which women express this sense of detachment are very vivid.

Since they had to put me to sleep, it was like it didn't happen. The doctor went over it all and gave me something to read, but it still seemed like something that happened to someone else. I would look at the baby and remember being pregnant, and there was nothing to help me get from pregnancy to baby in my mind.[34]

I didn't get to see the delivery. I wasn't even awake and my husband had to wait outside. Now I'm not even sure she's mine. Sometimes I really wonder. I have this dream where the nurse takes this baby out of the closet in the section room and gives her to me, and my baby is taken away. I'll never really know if she's mine.[35]

I had difficulty believing that I had, in fact, borne my seven-pound, six-ounce daughter. At times during those first few postpartum days, she seemed more like a doll handed to me by my obstetrician.[36]

My Lamaze teacher came in after the delivery. I could barely recognize her. I couldn't communicate with anybody. I was wondering where is my baby! Oh, I've missed everything! I wanted it to be different. At the same time I was sort of relieved. Oh, it's over. Then I saw Rebecca. I felt sort of irritated inside. Here I am the mother, and everybody had seen and felt her and here I am getting my chance for the first time hours later. The nurse helped me hold her. I was in pain. The whole time I'm thinking where have you been, you know, now I get to hold you and you are not new anymore, somebody's handled you and I wanted to be the first one. And of course that was all too late. (Janice Sanderson)

You read about how the maternal instinct is so strong, mother animals dragging young out of the forest fire. Maybe they would, but I was lying there feeling horrible and I couldn't have cared less. I drifted in and out. There was no "This is mine." It was objective, no feeling of attachment. That was scary, it created a lot of guilt. I had to cover it up because I didn't want the nurses to think I wasn't grateful for this child. I knew all the things you're supposed to do, the "en face" position and all that. It could have been a banana wrapped up in a blanket. (Laura Cromwell)

Most of the women's health literature explains this detachment as a result of the lack of opportunity for the mother and baby to "bond" because the mother's postsurgical condition and the baby's are usually thought to need special attention: mother goes to recovery and baby to be observed in the nursery.[37] The research on bonding is based on observations of specific behavior that mothers and babies display during the period immediately after birth, including touching, gazing, and smelling.[38] This behavior is often described as if it were automatically produced by proximity of mother and baby: an "instinctive," "programmed," "primeval dance."[39]

I would like to suggest that focus on the process of bonding that automatically takes place in the moments after birth (and efforts to restore this time even to cesarean mothers and babies) diverts our attention from how the whole process of birth has been sundered by the application of the production metaphor. To put it baldly, if a worker feels lack of involvement with a product when she does not feel she made it and when her work is regimented and controlled, think how a woman would feel if in her view her baby's birth was taken away from her by the exertion of control over her body? Surely restoring contact between mothers and babies immediately after birth could not restore automatically a sense of engrossment with the baby when the process of birth has been so deeply interrupted.[40]

I have stressed so far the dominant ways women talk about menstruation, menopause, pregnancy, and labor, all of which involve fragmentation of the self. Far less common, but occasionally present, are expressions—such as "*my* contractions," "*my* labor"—that depict the woman as actively experiencing the events. One woman objected to the prevailing attitude that pregnancy is something that she was afflicted by: "It's like all of a sudden some people would say you've got this condition. What condition? I'm pregnant, I don't have this condition" (Jan Moore). And one woman developed an image for thinking about her approaching labor that made it an act she would perform: "I'm hoping it will be a beautiful experience, almost like a blossom opening or something. It will be beautiful. This is my fruit, you know. It will be a wonderful experience because of the joy in it. The joy will take away from the pain and I'm willing to face the pain" (Pat Ladd).

But this image stands alone among the many others in which the woman is passively being *done to:* "like being spread from wall to wall wide open," "it felt like I imagine it would feel if someone was trying to pull your tongue out with pliers" (Jan Moore).

Occasionally, women move toward expressions that convey wholeness between themselves and menstruation, using active verbs or imagery of integration. One woman spoke of herself as a "menstruator," but only in the midst of many other passive constructions and images viewing her period as something from outside: "I always know when I'm going to get my period. I have none of the waiting, when you think you're around that time. I always know when it's going to happen. I was always a very light, controlled menstruator" (Mara Lenhart). And another struggled to find a way to see her period as part of herself: [How do you think you might feel about menopause?] "I haven't thought about it. In fact I usually don't think about having my period. It's just part of me. You don't think about having hands, you *have* your hands. I mean, my period is more than just like having hands. It's something more than that, but it's more second nature now. I'm not sure if it would be strange not having it, since it's not there every minute, every second. I can feel my hands wherever they are. Your period in a way is so infrequent, I guess it would be hard to miss, I'm not sure" (Kristin Lassiter). Still another succeeded in a clear assertion that menstruation was a part of herself: "It's not something that happens *to* you, it's not this outside thing that happens to you and that it's bad; it's something internal. There's a difference between something that happens outside of you and something that happens inside of you" (Carla Hessler).

The dominant verbal images in the interviews and birth reports describe a separation of the self and the body (the women's body sends her signals, labor is a state one goes through) which entails a separation between the self and physical sensations ("*the* contractions *come* on," etc.). When I realized the extent of these locutions in the interviews, I wondered by what route they had entered women's minds. We have seen how medical science has given us the notion that the uterus is an involuntary muscle. But it seemed unlikely that such a notion would be carried unchanged into the women's health movement or prepared childbirth texts and classes, which women are now widely exposed to. Surely, I thought, in feminist literature on health or Lamaze texts and prepared childbirth classes, menstruation, menopause, and birth would be described as acts a woman *did,* and physical sensations would be described as being intrinsically a part of her. When I looked at these texts, I did indeed find many statements that convey a sense of unity about the person and her physical experiences: Birth "is not something that you simply let happen to you . . . it is something that you *do*."[41] "Labor is called labor because it is the hardest work most people will ever do."[42]

However, alongside statements such as these I found statements of very different import. With respect to menstruation, an article that exhorts women to repossess their bodies, make menstruation their own, and redefine it positively for themselves also quotes women approvingly when they express the sort of fragmentation we have seen:

> My period, the fact that it makes me feel so vulnerable for a couple of days beforehand, reminds me that I have these vulnerable feelings and that I should stay in touch with them. I sometimes feel that it's my body's way of reminding me to stay sensitive and in touch with myself.

> The best way I can describe them [cramps] is to say I would be happy to lay down in bed with them, but we rarely get to do that together. Since we don't get the chance, my cramps and I, we do it okay in the world.[43]

When it comes to birth, alongside statements that birth is an act women do are statements that it is the uterine contractions that do the birthing: "Labor is the time when your uterus contracts to bring about the birth of your baby."[44] And more emphatically, uterine contractions are *involuntary:* "The uterus is an involuntary muscle, like the heart, and it is impossible to stop it from contracting."[45] Lamaze himself, who I had always thought stressed the ability of women to control the process of birthing, clearly separates areas of labor and delivery that a woman can control from those she cannot. On the one hand a woman is the "force which directs, controls, and regulates her labor"; on the other hand, her uterine contractions are "involuntary."[46] In labor the voluntary and involuntary actions come together: "pushed by the coupling of these combined effects—the one involuntary (uterine contraction), the other voluntary (mother's muscular efforts)—the child will make headway through the pelvic cavity."[47] That this fragmented imagery is found in the women's health movement literature is testimony not to poor intentions but to the invisibility and tenacity of these ideas and practices.

As we will discuss in more detail in the next chapter, in the nineteenth century women were seen as controlled in mind, body, and soul by the ovaries and the "paramount power" of the uterus;[48] the female was pictured as "driven by the tidal currents of her cyclical reproductive system."[49] Special care had to be taken when the uterus was active (during puberty, menstruation, and pregnancy) not to overtax it by mental or physical exertion. "Because reproduction was woman's grand purpose in life, doctors agreed that women had to concentrate all their energy downward toward the womb."[50]

Perhaps we no longer see women as so enslaved by their reproductive organs. But we are still a long way from seeing quintessentially female functions as acts women *do* with body, mind, and emotional states working together or at least affecting one another. The day has not come when instead of speaking of "hypotonic uterine dysfunction" or "nonprogressive labor" that has to be treated by chemical or operative intervention, we will say that "the woman has stopped her contractions" and focus on alleviating the fear and anxiety which have probably, in the majority of cases, led her to do so.

This chapter has shown us a fair amount of fragmentation and alienation in women's general conceptions of body and self. Ordinary women do not seem aware of the underlying fragmentation implied in the ways they speak and the assumptions they make about their selves and bodies; even women health activists, committed to struggle against medical models of health, fail to see these pervasive images. Although women resist specific medical procedures such as cesarean section or anesthesia during delivery, they seem unable to resist the underlying assumptions behind those procedures: that self and body are separate, that contractions are involuntary, that birth is production. The most resistance we see to this level of ideology is a kind of reluctance, a feeling that something is not right, and tortured efforts to reconcile experience with medical expectations. One woman I interviewed labored for about twelve hours at home and found it a "*really* nice experience." When she and her husband arrived at the hospital,

> I wasn't upset or anything and we got there and the obstetrician wasn't there and this young female resident examined me and just got freaked out, she got really nervous. She observed some blood and she felt that perhaps the placenta was down over the cervix and she just didn't know what was going on. She just scared [my husband] and me to death and she wouldn't let me walk and rushed us into this room next to the O.R. and it was freezing cold and they wouldn't give me a blanket. It was *very* upsetting.

Later a sonogram showed that the placenta was positioned normally; the resident's exam had produced the blood.

> It was a horrible experience; we were both so scared and from then on things just didn't progress terribly well. I didn't dilate over 4 cm after a few hours of pitocin so they decided to do a c-section. I was *real* disappointed.

Aside from the fear, certainly other factors affected the difficulty of this woman's labor after she got to the hospital: she was made to wear an oxygen mask, she was not allowed to stand or walk around, and she felt both her obstetrician and the labor room nurses could have been more supportive and sympathetic. But in her own perception of events, the fact that she was frightened to death was central:

> Immediately after the birth I thought that I got so scared and so uptight that my contractions sort of stopped for a while and then eventually started again. I thought maybe it had been something voluntary, for whatever reason.

But her perception that her fear may have led her to stop her labor is then immediately denied:

> But I read in the paper that there was a woman who was in a coma from a car accident and when it was time for the child to be born the child was born and she still is in a coma, so I guess it really is involuntary. Whether you're really upset just doesn't make any difference.

In her final statement, her inkling that having her baby might have been something she could *do* is paradoxically reasserted and then denied again. Finally, the fact that she did not accomplish the birth herself is put down to her own failure.

> I felt like my body had failed me. I really felt that having a baby is something you should be able to just *do*. And it's so much out of your control, I mean you can't really try harder, I felt sort of angry at myself. (Ellen Crowder)

The convolutions in this woman's interview are the result of a very real dilemma: if she takes seriously the possibility that childbirth is an act she as a whole person might have done, then the fact that she did not do this must be laid at the feet of the hospital, of herself, or both. Any of these possibilities would probably be more threatening than seeing childbirth as something that "happens to you" and labor as something that "really is involuntary." Adrienne Rich observes:

> I think now that my three unconscious deliveries . . . (approved and implemented by my physician) and my friends' exhilaration at having experienced and surmounted pain [through some version of natural childbirth] (approved and implemented by their physicians) had a common source: we were trying in our several ways to contain the expected female fate of passive suffering. None of us, I think, had much sense of being in any real command of the experience. Ignorant of our bodies, we were essentially nineteenth-century women as far as childbirth (and much else) was concerned.[51]

Fortunately, our story does not end on this distressing note. In the following chapters we will find other discourses existing alongside these, active assertion alongside passive acceptance, clear perception of alternatives alongside confused discontent, wholeness and integration alongside separation and fragmentation.

6 Menstruation, Work, and Class

I would just tell her that it comes at a certain time, and I would definitely inform her. A friend of mine said she thought that somebody put glass in her bed and she thought that she was cut and didn't tell anybody because she was always bringing in rocks. Figured she would get a beating. Went on for two years. She came on when she was nine. [What did she do with the sheets?] Threw them away. So happened her mother was taking trash out one day and seen them. Her mother says, "Why didn't you tell me?" She said, "I don't know how I keep getting glass in my bed." Her mother just laughed.
—Linda Matthews

We have seen that the image of menstruation in medical texts is unabashedly one of failed production. We have also seen that our general cultural conceptions place menstruation in the private realm of home and family. The task before us now is to see how women themselves—women in various life circumstances—regard menstruation, whether they have learned and adopted the medical model and the general cultural model or whether they have developed other kinds of models. In this chapter I discuss women who regard themselves as menstruating in an ordinary fashion, and in the next chapter women who regard themselves as having premenstrual syndrome, which is now considered a pathological malfunction.

Let us begin with the general cultural model. When women talk about menstruation they usually do not see it as a private function relegated to the sphere of home and family, but as inextricable from the rest of life at work and school. Accordingly, the way women talk about menstruation differs according to their expectations about the kind of work they will do outside the home, if any, and the nature of the career or further education, if any, they envision. These in turn differ markedly between the middle class and the working class.

The closest women come to talking about menstruation as a function that belongs in the private realm is when they describe it as "a hassle." What makes it a "hassle" has numerous dimensions. Women often see menstrual bleeding as "messy" and the blood itself as "gross" or disgusting. Some express feelings that menstrual blood is fearful and defiling. Many women mention fear of death or disease: "seeing all that blood," "lying in a bed full of blood," "I was sure I was bleeding to death," "saw that blood and thought I had TB," "thought I'd injured myself," "thought that I was dying." Others mention feeling dirty.[1] Most of these remarks relate to use of pads: young women recount their disgust at the blood on pads before they learned to insert a tampon, and older women recount their disgust at the blood on pads before tampons were available.[2] Others relate to messages from parents or friends that the process is unclean: older friends who routinely douched "made me feel like 'Oh, my God, I'm dirty or something!'" (Marcia Robbins). A mother "who had no education . . . thought it was dirty" (Ann Morrison). "People make you feel bad, especially men. They think that you are dirty when you have your period" (Gracie Evans). Many have tried to dispel these feelings about "the dirty secret": "something I've had to try to deal with, that I'm not dirty, that I don't stink and smell. It can work against you. You start feeling inadequate and irresponsible, just a dirt ball and you don't know why" (Ann Morrison). "The most important thing is to dispel that it's dirty or unclean. It shouldn't be this thing that makes women dirty and unclean" (Ellie Yamada).

But because women are aware that in our general cultural view menstruation is dirty, they are still stuck with the "hassle": most centrally no one must ever see you dealing with the mechanics of keeping up with the disgusting mess, and you must never fail to keep the disgusting mess from showing on your clothes, furniture, or the floor.[3] But problems arise precisely where menstruation does not belong, according to our cultural categories: in the realms of work and school outside the home. The "hassle" refers to the host of practical difficulties involved in getting through the day of menstruating, given the way our time and space are organized in schools and places of work. In high schools, how does a woman find time and private space to change pads or tampons so she won't "show"? "In school it's hard; teachers don't want to let you out of the classroom, they're upset if you're late and give you a hard time" (Kristin Lassiter). "In seventh grade I didn't carry a pocketbook or anything—wow!—How do you stash a maxi-pad in your notebook and try to get to the bathroom between classes to change? It was like a whole procedure, to make sure nobody saw, that none of the guys saw. From your notebook and into your pocket or take your whole notebook to the ladies' room which looks absolutely ridiculous" (Rachel Lehman). Strategies for dealing with these problems are many: put the maxi-pad up your sleeve, tuck it in a sweatshirt, or slip a tampon in a sock. One woman took "a year's supply of tampons" with her to the SAT's and rushed out during breaks (Lisa Miner).

If high schools are difficult places to obtain enough supplies while keeping their use out of sight, the same can surely be said of most workplaces. "I have to go to the bathroom more often and because I work with two men, I can't just say 'oops' and go" (Lisa Miner). One woman remembered an "excruciating incident" on her first day delivering goods to many small convenience stores, accompanied by her boss. She had incapacitating cramping in addition to needing frequent stops at a bathroom. "I couldn't even tell him, because that stuff isn't spoken of, what the matter was with me" (Meg O'Hara).

But the woman trying to sneak a tampon from the classroom into the bathroom and the woman who cannot tell her boss what is the matter are both being asked to do the impossible: conceal and control their bodily functions in institutions whose organization of time and space take little cognizance of them. But lest we see women as passive victims of an ideology that demeans them, we must ask whether women have been able to use any aspect of the shamefulness of menstruation to serve their own interests.

Let us begin with the bathroom, the place that women depend on to take care of their bodily functions including menstruation. I think it can be shown that this place is now and was in the last century a complex backstage area in contrast to the school, factory, or firm's public front stage area. Contemporary women students or workers can use bathrooms not only as places to keep their menstrual blood from showing but as places to preserve a bit of autonomy and room for themselves in a context where their physical movements are often rigidly controlled.[4] As one factory worker explained, "I was so tired all the time I was pregnant. There were no women supervisors and the men couldn't come into the john after me. I had an old coat I left in the john. I would put it on the floor and lie down. It was gross but you do what you have to do . . . If I asked for the right to do this, they'd tell me go jump in a lake."[5]

These private areas also allow certain forms of resistance: workers can take many breaks in the bathroom (when these are on the factory's time and do not detract from the workers' wages); while they are there, they can plot to avoid some management policy they dislike.[6] The double-edged nature of the shame of women's bodily functions here works in their favor: if private places must be provided to take care of what is shameful and disgusting, then those private places can be used in subversive ways.

Contemporary workers can probably take for granted the existence of more or less usable backstage areas designated for women. But in the nineteenth century this was not so. Accounts of working conditions in the nineteenth and early twentieth centuries make it clear that sanitary conditions could be appalling. Workers often ate their lunch on the workroom floor for want of a lunchroom or in dark and unventilated places.[7] (See Figures 20–22.) (The lunchroom pictured in Figure 23 was very much an exception.) If pleasant surroundings were not provided for lunch, even less were they provided for more private functions: toilets ranged from a "dirty little closet in the side of the wall"[8] to six holes cut into a plank over a river.[9] It is clear that these places were not pleasant retreats; to avoid them women sometimes changed clothes "with astounding frankness" on the workroom floor.[10]

These difficulties in finding private areas in public places faced by white women in early factories would have been faced even more acutely by black women whenever they moved in the public realm. Toni Morrison's *Sula* contains a fictional account (probably not far from the truth) of a train ride from Birmingham to New Orleans in which there were no public facilities for black women. In desperation the women used the grass alongside the station, under the eyes of white men.[11]

In industry, even when private rooms were available, it would not have been easy to use them for seditious purposes, for constraints on the workers' time could be extreme: washing hands, eating lunch, and going to the bathroom often had to be done during the half-hour break in a ten-hour day.[12] Especially when working on piece rates, women kept time off to a minimum. An observer overheard hoop skirt makers saying, "Let's swallow our dinner and when we have time chew it."[13] Beyond this, women might feel further constrained if they

Fig. 20 Women at work in a medical laboratory in Lowell, Massachusetts, in the late nineteenth century. (Penny 1870:396. Library of Congress.)

had to ask a male supervisor for permission to "go upstairs," a supervisor who would "look and laugh with the other clerks."[14] Upstairs they might well find a bathroom for the common use of both sexes but "many sensitive and shrinking girls have brought on severe illnesses arising solely from dread of running this gantlet [sic]."[15]

Fig. 21 The coil-winding room of the Westinghouse Electric and Manufacturing Company in Pittsburgh showing the conditions in factories in the early twentieth century. (Butler 1909, facing p. 219. Copyright © 1909 by the Russell Sage Foundation. Reprinted by permission of Basic Books, Inc., Publishers.)

In spite of these impediments, evidence shows that women still managed to use these backstage areas for solidarity and resistance. In early twentieth-century documents, there are scattered references to groups of two or three girls frequently found in the washroom "fussing over the universe"; groups of girls "exploding in the dressing room" because of a lack of soap and towels—"Gee! They sure treat you like dogs here"; a girl sobbing in the washroom over her stolen wages[16] and girls reading union leaflets posted in the washroom during a difficult struggle to organize a clothing factory.[17] Nor were managers insensitive to this threat. As recently as 1944 the Women's Bureau of the UAW passed a resolution complaining about company spies, who tried "to see if women were loitering about restrooms."[18]

Every taboo on something shameful has the potential for rebellion written in it: if what my body does must be kept secret, then I can use that opportunity to keep other things I do secret also. Taboos on women's activities while they menstruate cut two ways. In Anglo or European history, the taboos were based on beliefs that menstruating women cause meat to go bad, wine to turn, and bread dough to fall. In Cambridgeshire, well into this century, menstruating women could not touch milk, fresh meat, or pork being salted, lest it go bad.[19] Sim-

Fig. 22 Packing cigars in Pittsburgh in the early twentieth century. (Butler 1909, facing p. 75. Copyright © 1909 by the Russell Sage Foundation. Reprinted by permission of Basic Books, Inc., Publishers.)

ilarly, elsewhere in Europe, menstruating women could not touch wine or virtually anything else on the table: salted pork, a newly killed pig, salad dressing, mayonnaise, preserves, sauerkraut, pickles, or bread.[20] In a contemporary Welsh village, similar restrictions govern menopausal women.[21] These restrictions are usually put forth as evidence of the ways women have been kept from acting freely in society, as a result of men's horror of the products of the uterus.[22] I think an alternative account can be derived from the relationship between these taboos and the kind of work women were expected to do. In societies such as these, in which one of women's preeminent work duties was (and is) the preparation of food for the table (salting pork, salting vegetables, making bread, cooking meat and vegetables and serving it all), these taboos could very easily be perceived as providing a woman a welcome vacation.[23] After all, if she cannot touch the food to make her family's meal lest it go bad, she is automatically relieved of a large part of her daily tasks.

In other societies, similar "restrictions" are upheld by women. In China, a woman who has borne a child must stay in her room and eat only nourishing foods for a month afterward because she is "unclean" for that length of time. Although women chafe at the confinement,

Fig. 23 A lunch room for workers in a cigar factory in Pittsburgh in the early twentieth century. (Butler 1909, facing p. 311. Copyright © 1909 by the Russell Sage Foundation. Reprinted by permission of Basic Books, Inc., Publishers.)

they regard this period positively specifically because it is free of work. I was told many sad tales of desperately poor families who were unable to free a new mother from work for this long, and the dire consequences to her health that followed from her inability to regain her strength. If we are to describe a menstrual "taboo" as restrictive on women's behavior we must be sure to distinguish activities she longs for from those she is gladly rid of, at least for a time. We must also be careful to distinguish between domestic unwaged labor and waged labor, for when taboos like these are used to keep women out of whole occupations, such as dairying, winemaking, or silkmaking, the relationship between menstruation and work is very different.[24]

Although women easily recognize that menstruation and the world outside the home are incompatible, no women discussing the hassle and mess went so far as to ask why the outside world does not legitimate the functions of women's bodies.[25] Academics do not always manage this either. In her insightful study, Kristin Luker describes the very different relationships that pro- and anti-abortion women have to waged labor. She sees this split as a clash of worldviews, those on the one side valuing the contributions they make through their jobs in the outside society and setting reproduction aside, those on the other side valuing the contribution they make through their reproductive roles and setting outside production aside, in part because they are not able to get very rewarding jobs.[26]

Important as Luker's insights are, I think it is crucial to put these worldviews in a broader context of institutional power, asking why the workplace in this country is so incompatible with women's reproductive roles. Compared with other major industrial countries, the United States is extreme in the extent to which its institutions make productive and reproductive roles incompatible. In 1919 and again in 1952 the International Labor Organization (ILO) in Geneva drafted desirable standards for minimum maternity benefits for women workers, "including six weeks' mandatory leave after the birth of a child, six additional weeks' optional leave, a guarantee that a woman can reclaim her job at the end of the leave, and a cash allowance to assure healthy maintenance of the newborn infant." Although only eight European countries have adopted the ILO's specific conventions, most have guaranteed some minimum leave of absence with job security and cash benefits at some level.[27]

The United States is alone among all major industrial countries in its lack of a national insurance plan that pays medical expenses for childbirth. "It is one of the few governments in industrialized nations that does not provide any cash benefits to working women to compensate for lost earnings. Only in October 1978, after a two-year legislative battle, did the federal government enact legislation to guarantee that a working woman who bears a child is entitled to sick-leave benefits ordinarily provided by her employer."[28] This lack of institutional support in the United States makes it very difficult for women to be whole people—productive and reproductive at the same time. "Employers object to women's desire to have their cake and eat it, to reproduce and to go on earning . . . This desire is considered unrealistic, but in fact, the only way to be realistic in the world of work is to be a man."[29]

Luker puts the dilemma of the anti-abortion activist eloquently:

> Once an embryo is defined as a child and an abortion as the death of a person, almost everything else in a woman's life must "go on hold" during the course of her pregnancy: any attempt to gain "male" resources such as a job, an education, or other skills must be subordinated to her uniquely female responsibility of serving the needs of this newly conceived person.[30]

Of course, some women might define their needs while pregnant as all-consuming no matter what, but the current structure of workplaces in the United States does not easily allow any woman to live with her bodily functions, whether she be menstruating or pregnant.

If the shame and disgust attached to menstruation by the larger society sometimes allow women to escape their usual roles and act without the scrutiny of men, the positive feelings women themselves have about menstruation also allow them to act in their own behalf, not just as members of private families inside the home. The primary positive feeling many women have about menstruation is that it defines them as a woman. Part of the meaning of first menstruation is often a transition from girlhood to womanhood: "I remember wanting it so badly. You feel like you're not a woman but you know you're starting to be a woman and it's something that sets you apart from being a little girl" (Lisa Miner). "My mother said that I was becoming a woman, and that really got to me, so I said, 'Am I a woman now?'" (Kathleen Reardon). Mothers and sisters often greet the event with "You're a woman now!"

Sometimes the defining characteristic is closely linked with being able to have babies: "All women have it. I'm glad I'm female and that I have it because I can have babies" (Julie Morgan). But other times the defining characteristic is equally important apart from the potential to reproduce: "If a person doesn't want to have children, that's fine, but you still have to be a woman. I'm sure you could stop menstruation but I don't see any point" (Linda Ansell). Over and over again, women found different ways of saying, "No, I don't want to give it up, it's part of my self, part of what makes me a woman": "That's part of being a woman, you wouldn't want to get rid of it" (Mara Lenhart). "Sometimes I wish that I didn't have to have my period. I think everyone wishes that. But I've just gotten so much more attached to myself as a woman, it just seems so integral, I don't resent it anymore. If women didn't have to have periods, and nothing else would change, then that would be a wonderful thing. If someone said that to me individually, I'd have second thoughts. It would be nice to not have to get my period, because it's painful and changes my moods, but it seems necessary, it's supposed to happen. [Necessary?] Women get their periods and I'm a woman so I should get my period, that's what I mean by necessary" (Linda Ansell). "When I have it I think, I'm a woman, this is proof, that I get my period" (Shelly Levinson).

This common identity can often become the basis of common action. "It joins you together. It's like the one thing that all women have in common. It's great if your friend has a Midol" (Lisa Miner). Women who are relatively late starting menstruation feel acutely the difference it makes to be a part of a menstruating group:

> I remember girls used to always sit around and talk about having periods, whether you used tampons, teaching others to use tampons. Girls who had their periods longer taught girls who didn't use tampons how to use them. In junior high school, I felt so bad I didn't have it, I was left out from all this discussion. I had to keep quiet, I couldn't talk about it. So I felt when I got it, it allowed me to join these discussions and talk about using tampons, and getting cramps. It was all very important, all a part of being a junior in high school. (Shelly Levinson)

> The first time I had my period I was so happy that I remember singing in the bathroom, "You are a big girl now." I was so anxious to have my period because I always hung out with older kids and they had been through this and they had napkins and I always felt like such a baby. I would pretend and I would tell them that I was having my period now too. So when it finally happened I didn't have to pretend anymore. (Gina Billingsly)

Another woman ran to tell her friends in the school cafeteria that her period had just started: "They said, 'Wow, that's great,' and they bought me some ice cream and it was a really fun experience. When the last person in our group got her period we were all like a team" (Rachel Lehman).[31]

A part of feeling joined together as women is feeling different from all men. Anne Frank spoke of menstruation as her "sweet secret," and women I interviewed also spoke of it as a "secret," especially from men.[32] "I would walk around during the day and think that they didn't know, like a secret" (Shelly Levinson). Men were said to "have no concept of it" and often to completely misunderstand the process. Women found it hilarious to recount their discovery of these misunderstandings: a man who thought *all* women menstruated on the twenty-eighth day of the month because the calendar in his textbook had the twenty-eighth day marked and the man who thought all women menstruated for twenty-eight *days!*

Many women expressed the wish that men could menstruate for at least one cycle so they would know what it is like. For some this may come from the realization that when men do think about it they think it would be "horrible, the worst thing that could happen" (Tania Parrish) or disgusting. When women talk about the disgusting mess or the discomfort, they do so with an implicit, often unstated understanding that there is another side to the process: it is part of what defines one as a woman, and it is something all women share, even if what we share is talking about the problem of dealing with this disgusting mess. When women hear men say it's horrible or disgusting, they know full well that men are in no way appreciating the other side. So wishing they could experience it may be a way of wishing they could consider the process in totality, not in part.

Descriptions of menstruation as a "hassle" and as a mark of womanhood stretch across all interviews. Other aspects differ markedly by class. It is clear that women construct the significance of menstruation in terms of the range of opportunities open to them and their expectations about how they will make use of them. In response to the question "How do you think you'd feel about being pregnant?" a middle-class teenager said, "If tomorrow, I'd be real scared. I've kinda thought about that, I can't say what exactly I'd do. I'd be extremely scared. I would talk to my sister about it, my mother. I'd consider abortion, I'd be real scared" (Kristin Lassiter). Other women, older but also middle class, express their determination to menstruate for years before pregnancy. "I know I personally would want to have children, I don't really have fears about it, I just don't, unless it happened to me tomorrow, which it won't (Linda Ansell). "In the short term I would be fearful of being pregnant. I suppose at some point in my life I'll want to have children. I don't know when, it's not easy to say. I want to be firmly established. I don't know how I will balance holding a career and having a family. If I became pregnant now I'd have an abortion, but I would hope not to have to" (Mara Lenhart).

In contrast to these remarks by middle-class high school and college students are the accounts of working-class women in a high school specifically for previous dropouts, many of whom already have a child. Although these women are oriented toward college and a career, the colleges they enter will for the most part be local state universities or community colleges, the careers they choose second choices that take into account the lack of family resources: "I think I'd be a good lawyer, but I'd like to work with children, probably as a social worker" (Lisa Stephens).

Some of these girls mention their career plans when asked about pregnancy, just as the middle-class girls do: [What kind of birth experience would you want?] "It all depends on how my life would be five, six years from now. It all depends on the future" (Sandy Hammond). [Overall how do you feel about menstruating?] "At times I feel it's the best thing that's happening. If I'm coming on my period it means I'm not pregnant. I'm glad to see it. [Do you want to have children?] Yeah, I think after I get my career going. When I can take care of myself, that's when I want a child, when I can take care of him and me" (Kathy Spencer). "I do not and I will not get pregnant at a young age. I want to finish school and make something of myself and have the money to support myself and a child before I do" (Rose Oliver).

For most, a threat to the regular appearance of menstruation or the direct prospect of pregnancy does elicit fear, but only about the event of birth itself.[33] [Any hopes or fears about pregnancy?] "No, I wish I could have a little girl, but I don't want to go through that pain. [If you got pregnant now, what kind of experience would you want to have?] Natural childbirth. I would want somebody [the father] to be in pain too, not just me alone!" (Mary King). [Any hopes and fears about pregnancy?] "I think the only fear I would have is having it. I have heard that it is a nice feeling, but the other side is painful, very painful. It would be nice once it comes out" (Sandy Hammond). And for others, pregnancy is simply desired: "My mother wants me to wait until I am in my twenties to have a child, but I want one now. I want one while I am still young. I don't want to be an old fogy, and be forty when my kid is sixteen. Like my mother, she is 46 and I am 17. I want me and my child to grow up together" (Cathy Roark).

Recent studies have given us important insights into the reasons why childbearing has a different significance for working-class black and white women than it does among the middle class. Carol Stack's study of a poor, black, urban neighborhood stresses the extent to which child begetting is desired as a positive good and the ways in which resources are pooled to care for those children through a variety of overlapping sharing links among kin and households.[34] Joyce Ladner has shown how, in the absence of many other resources, sex comes to be seen as a resource poor black teenagers can exchange,[35] and Lillian Rubin has made plain how pregnancy, which girls hope will be followed by marriage, leads to one of the very few kinds of "mobility" available to young, white, working-class girls, enabling them to move out of their natal households and establish their own.[36]

It is hard to miss the implication that where access to the forces and means of production in the society is severely limited, reproduction can be (or can at least appear to be) a way of acting in the world, changing one's life, producing a "resource" to be shared or cherished.[37] In her research, Kristin Luker found that the break between pro-choice and anti-abortion activists ran along these lines: pro-choice women regard their fertility as a handicap to their productive roles;

anti-abortion women, because productive roles are less able to be rewarding to them, regard their reproductivity as a resource to which other roles must be made secondary.[38] Although most of the activists Luker interviewed were white middle-class women, Ladner has pointed out that poor blacks and the new right share moral disapproval of abortion.[39] It seems to me that at least the two extreme ends of the spectrum could be described as pro-production versus pro-reproduction, precisely because of the different relationship women at the two extremes have to the rewards of productive work.

Given this, an interesting question arises. The view of menstruation promulgated by scientific medicine is that it is failed (re)production. Do women whose stake in reproduction differs a lot, along the lines suggested above, also differ in how comfortable they feel about this view of menstruation or how willing they are to believe it? Among middle-class women, both black and white, two questions in the interviews elicited a version of the scientific view of menstruation in every case. The questions were: "What is your own understanding of menstruation?" and "How would you explain menstruation to a young girl who didn't know?" Accounts differed in details, but almost all began immediately with internal organs, structures, and functions.

> When you reach puberty, hormones in your body cause your—you have eggs in your ovaries—I'd explain the structures—the hormones in your body cause some of these eggs to ripen and they are released into your fallopian tubes and this is where pregnancy would occur. Your uterus has prepared the lining in case of fertilization and if there is not fertilization then the lining is sloughed off. It comes out in a bloody sort of tissuey substance. (Tania Parrish)

> Using the book in front of us, there is the cycle and the pictures. With the egg coming out and going in to nest at the vaginal wall and when it hasn't been, I can't think of the word, by the sperm, it would come out in the form of blood. And I found that very interesting. Waste matter with the blood. I think of it as just blood. The nesting place that is no longer needed. (Margaret Crichton)

Almost all who spoke of internal organs conveyed the sense that the purpose of the whole process is to provide key steps in reproduction, and that if menstruation occurs it is a waste product of an effort that failed: "All the blood that comes out with the egg was sort of like a home for the egg, so if the egg doesn't have to grow bigger into a baby you just say that it leaves with all that" (Mara Lenhart). "If the egg isn't fertilized there isn't any need for that lining, so the egg leaves the uterus without being fertilized and essentially dies. If the egg isn't implanted then that whole lining comes away and that's the menstruation; it leaves the uterus; it decays and leaves the uterus through the vaginal opening and that's what the period is" (Carolyn Potter). "If you are not pregnant each month your body has to get rid of the waste which is the eggs that are released from the fallopian tubes" (Eileen Windell). The central metaphor of failed production of a baby comes through vividly.

Other middle-class women (particularly high school age women who have heard this story a lot less often than older women) became somewhat uncomfortable in the interview when they couldn't remember to their satisfaction just what body parts were doing what: "Isn't it one egg from each tube every month and it changes, something like that, I don't remember, I don't know all about the stuff that's going on technically" (Julie Morgan). "I'm bad at explaining things. A process of an unfertilized egg, making its way through your fallopian tubes, right? Into the ovary, is it? It's not—it embeds itself into the walls of the ovary and as it doesn't become fertilized, the walls of the ovary which have—I can picture it all but I can't explain it" (Kristin Lassiter). In spite of these hesitations, all of them managed to get out some version of the failed production view.

Later in the interview, many of these women made remarks in passing about how this internal model was not relevant to them.[40] "When I have my period I don't really think about what's going on inside. I think about what's coming out of me. I don't think, 'The eggs are coming down.' It's just like I have my period and it's gone. Not like in the middle of the month and then: it's about time an egg is making its way into my fallopian tubes—I mean—!" (Kristin Lassiter). "The boys and the girls were separated and we all saw this film, which was so plain and dry in describing the process that you would think that it wasn't going on inside your own body. I thought that it was pretty pointless" (Carla Hessler). "I know all about the exact hormonal changes, but it doesn't really matter to me. Exactly when my luteinizing hormone takes over that part of the cycle isn't as important" (Carolyn Potter).

But despite this kind of dissatisfaction with the medical model, the interviews with middle-class women do not contain more than a glimmer of a different view of menstruation. Only one woman began her answer in a different mode, with what might be called the phenomenology of menstruation:

> [How would you explain menstruation to a young girl who knew nothing about it?] What happens in menstruation is that something that looks like blood but isn't in fact because it's darker comes out of your vagina and it—you—it has to be kept clean because otherwise you make a mess. What comes out changes its appearance. You have a flow coming out of your body for anywhere from three to seven days, and it changes. Sometimes it comes out very heavily, you can almost feel it coming out and sometimes it's very light and sometimes it stops during that time. And it changes consistency during the time it comes out. It looks very red and bloody, sometimes it's darker, sometimes it's almost brown. (Shelly Levinson)

This woman proceeded directly to the scientific model, saying that the process is caused by an egg that has died coming out with its fluids. The disjunction between this woman's two versions of menstruation cannot be overemphasized. In her first version, she deals with what menstruation feels like, looks like, smells like, what the immediate experience of being a "menstruator" is like, aspects of the process that are untouched by the medical model.

Even women who describe the scientific view of menstruation and say it is satisfying in one context find themselves at a loss with what it leaves out in another. One woman was given a "pretty good explanation" of "the mechanics" by a friend, her mother, and a fifth-grade sex education class. But when she started menstruating, she had no idea what to do when she went to bed. "I didn't know how you are supposed to cope with it. You change during the day, how could you sleep all night?" (Shelly Levinson). She was very nervous about asking her mother but when she eventually did she was told the flow slows down at night.

This same woman described a "terrible thing that happened" the first time she tried to put in a tampon. She got it in halfway, thought she had to leave it there and "went through a lot of agony that day." Other women struggle desperately to get the practical knowledge needed to insert a tampon (where to insert it, how far, what to do with the applicator) and can make painful mistakes, like leaving the applicator in with the tampon. The few women whose mothers, sisters, or friends take them in the bathroom (or, as in one case, shout through the door) and show them in concrete detail how to do this thing, are moved by the experience: they feel close to the person, indebted to them, and dedicated to teaching others. Aside from the difficult first rite of inserting a tampon, the middle-class interviews are full of anguish surrounding first periods, when women wonder Is the color of my blood all right? What do I wear? What if I skip a month? How often do I change pads or tampons?

If middle-class women readily incline toward the medical view of menstruation, even though this leads to difficulties dealing with the actualities of menstruating, how do working-class women explain menstruation? There the middle-class pattern is reversed: almost nobody—black or white—came out spontaneously with the failed production model of menstruation, and almost everybody accounted for it either phenomenologically or in terms of a life change. Here are some examples of working-class responses to the question "How would you explain menstruation to a young girl who didn't know anything about it?"

I'd just tell her it's when your body's changing and you're ready to have children. Your body allows you to have children now and it's one of the first steps to becoming a woman. (Valerie Bartson)

Your body's changing. [Is there anything else other than your body's changing?] No. [Your body's changing so does that mean anything to you?] Yeah, well, I think it's when you're on your period; I don't know really. All I know is your body is changing an awful lot. That's all I really know. [Later in the interview the interviewer returns to the same question.] I would say, well, you're gonna get on your period when you are about from nine to twelve and it's like an egg passing through you. It's . . . I couldn't even explain it to myself if I tried to! (Kathy Spencer)

I guess it's a part of your life, growing up. [Could you give a description of menstruation itself?] Yeah. [What would you tell her?] Just red blood. (Patricia Henderson)

I know, right. OK. [Doesn't have to be scientific, just your own understanding.] OK. It comes once a month usually sometimes five to seven days. I know I don't like it. You want to know why it came about or . . . [Just your own feeling of how it comes about or why.] Oh! I don't know really why it came about, I know my mother says if Eve didn't do this then we wouldn't have all these troubles with childbirth and menstruation and stuff like that. That's mostly my understanding, back to Genesis in the Bible, right? (Lisa Stephens)

I would tell her that her menstrual is like one of the worst nightmares. I don't know. I would tell her that it is something that she had to go through. With it comes displeasure because one, your period makes me sick because the blood, it ain't got the best odor in the world and I would tell her to check with the pharmacist for the best thing to use. Because tampons can give you toxic syndrome or something. But the Kotex is very uncomfortable. It is like a big bulk. And you feel it close to you and it is an icky feeling. Just like having sex and not going to wash up. And if she has sex she could get pregnant if she wasn't using some kind of protection. (Crystal Scott)

I don't know, it's part of mother nature I guess. I guess I would explain it the same way my mother explained it to me. [Which was?] It's just part of life, your body's changing and you're becoming a woman. (Kathleen Reardon)

I guess to clean out the body. [If there was a little girl who didn't know anything about menstruation and you had to explain it to her, what would you tell her?] I don't know. [Just picture me as being a little kid . . . (Interviewer proceeds to act out the role of a girl asking for information about a friend who had blood in her pants.)] I'd just tell her she came on her period. It's a cycle you go through every month. [What happens?] You bleed. [Why do I bleed?] To clean out your insides. (Jody Kelsey)

I would just tell her there's going to be a time when you're going to start seeing this blood and it's called your menstrual period. Don't be frightened or anything like that, because it's just a sign that you're growing up and becoming a woman. (Cheryl Dinton)

That it is nothing to be afraid of. That was something my mother told me. It is nothing to be afraid of. It is natural, and you have to live with it whether you want to or not. And just keep yourself clean. Make sure you change your pad or whatever you are using . . . [How would you explain what was going on?] Everyone says you are becoming a woman. That is not true, I don't know. You are thirteen years old and you get your period. And it is just something that you have to worry about. If you are fooling around with a guy, I would definitely tell him to make sure he uses something. Because if you miss your period you are in trouble. I guess you are changing into womanhood. It would be hard to explain. [What would you tell her about what was happening inside her body?] I don't know, it is hard to explain. For me I have trouble, you know, trying to explain something. I have trouble putting words together, or trying to think of the right words. (Mary Jo Lasley)

These women share an absolute reluctance to give the medical view of menstruation. This is so in spite of the interviewers' many efforts to give them a variety of opportunities to come up with it and in spite of the fact that all of them have been exposed to it in classes in school. In fact, many of the women mentioned how much attention was paid to menstruation in school: "had films and everything, books and pamphlets given out." In spite of this only two of the working-class women we interviewed mentioned anything inside the body: "the eggs passing through" and "to clean out your insides." All other responses involved only what a woman sees and feels or the significance it has in her life.

There are many possible explanations of this striking difference between middle-class and working-class women. Some might suggest that the difference occurs simply because working-class women are less exposed to scientific vocabulary, less familiar with the kind of written materials that schools use to disseminate the scientific model, or less able to produce the cultural performance of explaining something in an interview. Yet the interviews make it plain that these women can explain other things with ease and comfort and that they can integrate knowledge gained from books, say about a career they want to pursue.

It strikes me as at least equally likely that these women have simply been more able to resist one aspect of the hegemonic scientific view of women's bodies because it is not meaningful to them or because it is downright offensive, phrased as it is in the negative terms we have seen. The ironies in this are many: middle-class women, much more likely to benefit from investment in the productive system, have swallowed a view of their reproductive systems which sees menstruation as failed production and as divorced from women's own experience. Working-class women, perhaps because they have less to gain from productive labor in the society, have rejected the application of models of production to their bodies. It is striking that in the working-class interviews there is none of the angst that runs through the heart of middle-class interviews: almost no one describes being mystified at the details of menstruating (though they may dislike it), at a loss about the mechanics of taking care of it, or the variety of forms menstruating can take. The working-class woman who told the story in the epigraph for this chapter found her friend's ignorance bizarre and inexplicable. Mothers, grandmothers, sisters, and friends give these women the detailed information and practical knowledge they need: "My mother just told me how to use everything and that was it" (Kelly Morse). "My mother talked for hours the whole day about it" (Mary King).

Now that we have looked at how women talk about menstruation, how do we size up their overall consciousness of scientific and general cultural models? To be sure, there are many instances where women produce the same kind of oppositions between the valued public world of work and the devalued private world of family we saw in Part One. One woman said, "There should be celebration around menstruation. If done right it could be wonderful! But what would you be celebrating? The ability to bear children? You know there's that whole uselessness part of it, too, if you decide not to bear children" (Meg O'Hara).

But there are other instances in which menstruation is seen as an emblem of women's particular experience of the world, which begins to look very different from the cultural oppositions we laid out in Part One. Even though women do not often imagine it could be otherwise, they experience a reality on a monthly basis that confounds the separate spheres of men and women. They are also able to use this "anomalous" experience as the basis for common action by sharing its griefs, trading equipment, combining knowledge, and using the private spaces necessarily provided in school or work for refuge or rebellion. Beyond this, a portion of these women, working-class women, have a view of this process that runs exactly counter to the scientific model. Here we can learn from each other. Perhaps if we all gave our daughters a phenomenological as well as a medical explanation of menstruation, going into all the little details about how it feels and what you do, maybe we could restore a feeling of wholeness about the process and reduce in and of itself some of the disgust.[41] It may be that part of the impetus behind the alienated language we began with—it happens to me, I go through it, I get it, I am on it—is that women do not feel like menstruation is something they actually do. Articulating the sensory and emotional experience each person has may be a step toward that end.

Over all, middle-class women appear much more "mystified" by general cultural models than working-class women. They have bought the teleological aspect of medical accounts, which sees menstrual flow as waste products of a failed pregnancy and casts it in very negative terms. Within some of the interviews are faint protests: the "uselessness" referred to by Meg O'Hara, the "blindness about the body," or the "ignorance of the uterus" referred to by others. One effect of a teleological functional model is that it allows a set of internal organs to have only one purpose. Yet the woman who wants to bear children, the woman who is a lesbian and does not have sex with men, and the woman who does not want to bear children have vastly different relationships to the potential functions of those organs. By focusing on birth, a function which all female reproductive organs are supposed to fulfill, the medical model and women who accept it tend to conceal the true unity women have: we menstruate or have menstruated in the past. By making pregnancy the end point for which the system strives (and menstruation a wasteful failure), we come to despise menstruation—the one thing we all share, fertile, infertile, heterosexual, homosexual.[42] Ironically, in our society the great majority of the time most women are not intending to get pregnant. Most of the time, therefore, the arrival of menstruation could be seen as a welcome sign.

Perhaps we can escape the prevailing cultural view that internal invisible organs and structures in the body are central to menstruation, as the working-class women I interviewed have done. Or perhaps we could reappropriate the functions of those organs to our own ends. Maybe they developed that way because of the exigencies of our natural history, but it need not be the natural history of the species that determines the meaning of those organs and their processes today. If the body really is blind and the uterus ignorant (that is, not possessed of purpose), then—depending on our own purposes—they can just as well be making the life stuff that marks us as women or heralding our nonpregnant state as they can be casting off the hemorrhage of necrotic blood vessels and the debris of endometrial decay.

7 Premenstrual Syndrome, Work Discipline, and Anger

There are so many roots to the tree of anger that sometimes the branches shatter before they bear.
—Audre Lorde
"Who Said It Was Simple," in Chosen Poems Old and New, 1982

I turn now from what is considered normal menstruation to what is considered abnormal.[1] Looming over the whole current scene in England and the United States is the enormous outpouring of interest—the publishing of magazine and newspaper articles, popular books and pamphlets, the opening of clinics, the marketing of remedies—devoted to premenstrual syndrome.

The dominant model for premenstrual syndrome (PMS) is the physiological/medical model. In this model, PMS manifests itself as a variety of physical, emotional, and behavioral "symptoms" which women "suffer." The list of such symptoms varies but is uniformly negative, and indeed worthy of the term "suffer." Judy Lever's list in her popular handbook serves as an example (Table 1).[2]

The syndrome of which this list is a manifestation is a "genuine illness,"[3] a "real physical problem"[4] whose cause is at base a physical one.[5] Although psychological factors may be involved as a symptom, or even as one cause, "the root cause of PMT [short for premenstrual tension, another term for PMS], no matter how it was originally triggered, is physical and can be treated."[6] This physical cause comes from "a malfunction in the production of hormones during the menstrual cycle, in particular the female hormone, progesterone. This upsets the normal working of the menstrual cycle and produces the un-

Table 1 A LIST OF THE SYMPTOMS OF PREMENSTRUAL SYNDROME FROM A POPULAR HANDBOOK

Complete Checklist of Symptoms		
Physical Changes		
Weight gain	Epilepsy	Spontaneous bruising
Skin disorders	Dizziness, faintness	Headache, migraine
Painful breasts	Cold sweats	Backache
Swelling	Nausea, sickness	General aches and pains
Eye diseases	Hot flashes	
Asthma	Blurring vision	
Concentration		
Sleeplessness	Lowered judgment	Lack of coordination
Forgetfulness	Difficulty concentrating	
Confusion	Accidents	
Behavior Changes		
Lowered school or work performance	Avoid social activities	Drinking too much alcohol
Lethargy	Decreased efficiency	Taking too many pills
	Food cravings	
Mood Changes		
Mood swings	Restlessness	Tension
Crying, depression	Irritability	Loss of sex drive
Anxiety	Aggression	

Source: Judy Lever and Michael G. Brush, *Pre-menstrual Tension,* © 1981 by Bantam Books; reproduced by permission.

pleasant symptoms of PMT." Astonishingly, we are told that "more than three quarters of all women suffer from symptoms of PMT."[7] In other words, a clear majority of all women are afflicted with a physically abnormal hormonal cycle.

Various "treatments" are described that can compensate a woman for her lack of progesterone or her excess of estrogen or prolactin. Among the benefits of this approach are that psychological and physical states that many women experience as extremely distressing or painful can be alleviated, a problem that had no name or known cause can be named and grasped, and some of the blaming of women for their premenstrual condition by both doctors and family members can be stopped. It seems probable that this view of PMS has led to an improvement from the common dismissals "it's all in your mind," "grin and bear it," or "pull yourself together." Yet, entailed also in this view of PMS are a series of assumptions about the nature of time and of society and about the necessary roles of women and men.

Let us begin by returning to the nineteenth century, when, as we saw in Chapter 3, menstruation began to be regarded as a pathological process. Because of ideas prevailing among doctors that a woman's reproductive organs held complete sway over her between puberty and menopause, women were warned not to divert needed energy away from the uterus and ovaries. In puberty, especially, the limited amount of energy in a woman's body was essential for the proper development of her female organs.

> Indeed physicians routinely used this energy theory to sanction attacks upon any behavior they considered unfeminine; education, factory work, religious or charitable activities, indeed virtually any interests outside the home during puberty were deplored.[8]

The attacks must have been difficult to withstand. One doctor constructed a dialogue with a mother, who brought her fifteen-year-old daughter to see him, dull and moping, pale and thin. He urged the mother to take the daughter out of school, keep her home, and teach her "domestic administration." "*I should much prefer to have a daughter healthy, sweet-tempered, sensible, and beautiful, without Latin, and algebra, and grammar, than to have one ever so advanced in her humanities, with her health ruined, or, perhaps, lying under a marble urn at Laurel Hill.*" The hypothetical mother replied, "Why, doctor, you shock me!" and was told, "I wish to shock you; I wish you to learn that, unless you change the treatment, you will lose her. She will die, madam!"[9]

This view of women's limited energies ran very quickly up against one of the realities of nineteenth-century America: many young girls and women worked exceedingly long and arduous hours in factories, shops, and other people's homes. The "cult of invalidism" with its months and even years of inactivity and bed rest, which was urged on upper-class women, was manifestly not possible for the poor. This contradiction was resolved in numerous ways: by detailing the "weakness, degeneration, and disease" suffered by female clerks and operatives who "strive to emulate the males by unremitting labor"[10] while callously disregarding the very poor health conditions of those workers;[11] or by focusing on the greater toll that brain work, as opposed to manual work, was thought to take on female bodies. According to

Edward Clarke's influential *Sex in Education* (1873), female operatives suffer less than schoolgirls because they "work their brain less . . . Hence they have stronger bodies, a reproductive apparatus more normally constructed, and a catamenial function less readily disturbed by effort, than their student sisters."[12]

If men like Clarke were trying to argue that women (except working-class women) should stay home because of their bodily functions, feminists were trying to show how women could function in the world outside the home in spite of their bodily functions. Indeed, it is conceivable that the opinions of Clarke and others were in the first place a response to the threat posed by the first wave of feminism. Feminists intended to prove that the disciplined, efficient tasks required in the workplace in industrial society could be done by women when they were menstruating as well as when they were not.

In *Functional Periodicity: An Experimental Study of the Mental and Motor Abilities of Women During Menstruation* (1914), Leta Hollingworth had 24 women (who ironically held research and academic jobs) perform various tests of motor efficiency and controlled association both when they were and when they were not menstruating. These included tapping on a brass plate as many times as possible within a brief time; holding a 2.5-mm rod as steady as possible within a 6-mm hole while trying not to let it touch the edges; naming a series of colors as quickly as possible; and naming a series of opposites as quickly as possible. Ability to learn a new skill was also tested by teaching the subjects to *type* and recording their progress while menstruating and not. The findings: "the records of all the women here studied *agree* in supporting the negative conclusion here presented. None of them shows a characteristic inefficiency in the traits here tested at menstrual periods."[13]

Similarly, in *The Question of Rest for Women During Menstruation* (1877), Mary Putnam Jacobi showed that "women do work better, and with much greater safety to health when their work is frequently intermitted; but that those intermittences should be at short intervals and lasting a short time, not at long intervals and lasting longer. Finally that they are required at all times, and have no special reference to the period of the menstrual flow."[14] Given the nature of the organization of work, men would probably also work better if they had frequent short breaks. What is being exposed in these early studies, in addition to the nature of women's capacities, is the nature of the work process they are subjected to.

It is obvious that the relationship between menstruation and women's capacity to work was a central issue in the nineteenth century. When the focus shifted from menstruation itself to include the few days before menstruation, whether women could work outside the home was still a key issue. It is generally acknowledged that the first person to name and describe the symptoms of premenstrual syndrome was Robert T. Frank in 1931.[15] Two aspects of Frank's discussion of what he called "premenstrual tension" deserve careful attention. The first is that he carried forward the idea, which flourished in the nineteenth century, that women were swayed by the tides of their ovaries. A woman's ovaries were known to produce female sex hormones, and these were the culprit behind premenstrual tension. His remedy was simple and to the point: "It was decided to tone down the ovarian activity by roentgen [x-ray] treatment directed against the ovaries."[16]

Frank reserved x-ray treatment for the most severe cases, but it was not long before the influence of female hormones on a woman was extended to include her emotional states all month long. In an extraordinary study in 1939, Benedek and Rubenstein analyzed psychoanalytic material from therapy sessions and dreams on the one hand and basal body temperature and vaginal smears on the other from nineteen patients under treatment for various neurotic disturbances. They were able to predict when the patients had ovulated and menstruated from the physiological record as well as from the psychoanalytic record. Examples of their charts correlating hormones and emotions are shown in Figure 24 and Table 2. But Benedek and Rubenstein went far beyond showing simple correlations between hormones and emotions. Without evidence of a causal link one way or the other, they concluded that "human [they meant 'adult female'] instinctual drives are *controlled by* gonadal hormone production" (emphasis added).[17]

Benedek and Rubenstein may have been unusually avid in their attempt to derive women's emotional states from hormones. But their study was still being quoted approvingly and elaborated on in the late 1960s. One later study (1968) concluded that "the menstrual cycle exercises gross influences on female behavior."[18] It was not until the 1970s that some researchers began to insist that women's moods had important social, cultural, and symbolic components and that even though *correlation* between biochemical substances and emotional changes can be observed, "the direction of causality is still unclear. Indeed, there is abundant evidence to suggest that biochemical changes occur in *response* to socially mediated emotional changes."[19]

The second aspect of Frank's study that deserves attention is his immediate interest in the effect of premenstrual tension on a woman's

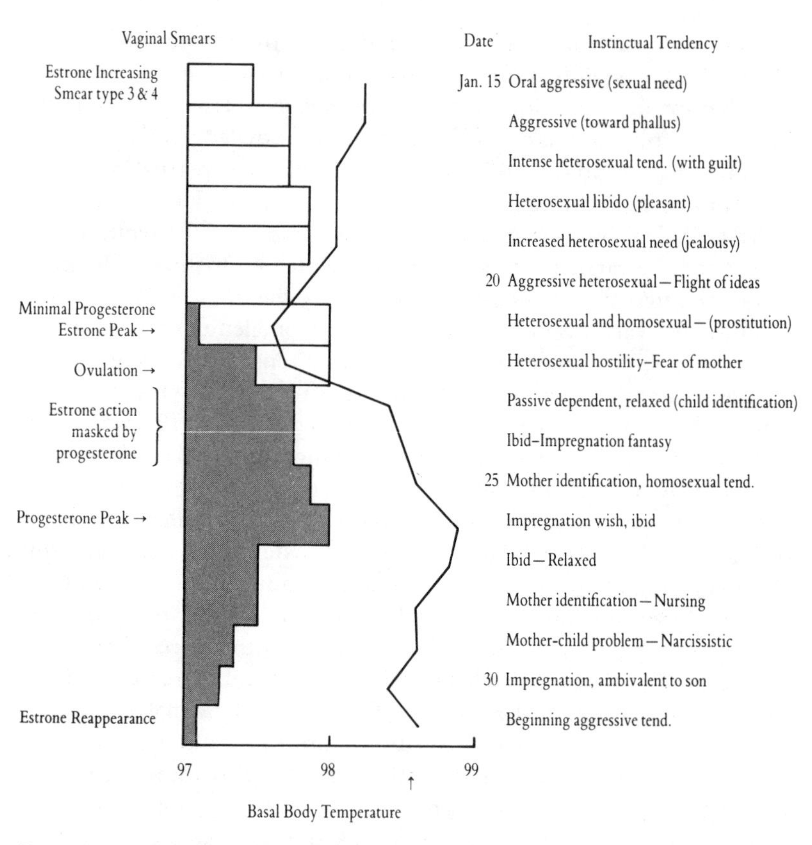

Fig. 24 Benedek and Rubenstein's correlations between hormonal production and psychological states in one patient. (Benedek and Rubenstein, *Psychosomatic Medicine,* vol. 1, no. 2, 1939, p. 268. Reprinted by permission of the National Academy of Sciences, Washington, D.C.)

ability to work, such that in mild cases employers have to make provision for an employee's temporary care and in severe ones to allow her to rest in bed for one or two days.[20] It strikes me as exceedingly significant that Frank was writing immediately after the Depression, at a time when the gains women had made in the paid labor market because of World War I were slipping away. Pressure was placed on women from many sides to give up waged work and allow men to take the jobs.[21]

Can it be accidental that many other studies were published during the interwar years that showed the debilitating effects of menstruation

Table 2 BENEDEK AND RUBENSTEIN'S GENERAL CORRELATIONS BETWEEN OVARIAN ACTIVITY AND PSYCHODYNAMIC PROCESSES

Diagram I

Hormone	*Instinctual Tendency*	*Neurotic Elaborations of Tendency*
Oestrone Follicular hormone	Active object libido on genital level: heterosexual desire	1. Aggressive incorporative: penis envy, castration wish 2. Masochistic: masochistic concept of female sexuality 3. Defense reactions: a) fear of being attacked b) masculine protest
Progesterone dominant Corpus luteum hormone	Passive receptive tendency on genital level desire to be loved: and wish for impregnation:	Passive receptive tendency on regressive level: oral receptive and oral dependent wishes, may be directed toward a) mother b) homosexual object c) heterosexual object

Diagram II

Phase of Cycle	*Hormone State*	*Psychological Material*
Follicle ripening	Initial oestrone function	Heterosexual tendency, usually pleasant, feeling of well-being
Late pre-ovulative	Increasing oestrone plus minimal progesterone	Relief by sexual gratification or increasing tension—conflicting tendencies (See Diagram I.)
Ovulative (immediately after ovulation)	Diminishing oestrone plus increasing progesterone	Relaxation of conflict tension. Erotization of female body, passive-receptive. Pleasant emotional state.
Post-ovulative, luteal	Progesterone dominance	See Diagram I, especially passive receptive tendencies and object libido toward mother or homosexual object.
Late luteal, early premenstrual	Diminishing progesterone plus resultant reappearance (unmasking) of oestrone effects	Recurrence of heterosexual tendency on mostly receptive level, and pregnancy fantasies

Source: Benedek and Rubenstein, *Psychosomatic Medicine,* vol. 1, no. 2, 1939. Reprinted by permission of the National Academy of Sciences, Washington, D.C.

on women?[22] Given this pattern of research finding women debilitated by menstruation when they pose an obstacle to full employment for men, it is hardly surprising that after the start of World War II a rash of studies found that menstruation was not a liability after all.[23] "Any activity that may be performed with impunity at other times may be performed with equal impunity during menstruation," wrote Seward in 1944, reversing her own earlier finding in 1934 that menstruation was a debility. Some of the evidence amassed for this conclusion seems astoundingly ad hoc: Seward argues that when women miss work because of menstrual complaints they are indulging in "a bit of socially acceptable malingering by taking advantage of the popular stereotype of menstrual incapacitation." Her evidence? That when a large life insurance company discontinued pay for menstrual absenteeism after a limited time allowance, menstrual absenteeism markedly declined![24] She missed the point that if people need their wages and have used up their sick leave, they will go to work, even in considerable discomfort.

After World War II, just as after World War I, women were displaced from many of the paid jobs they had taken on.[25] The pattern seems almost too obvious to have been overlooked so long, but as we know there was a spate of menstrual research after the Second World War that found, just as after the first, that women were indeed disabled by their hormones. Research done by Katherina Dalton in the 1940s was published in the *British Medical Journal* in 1953,[26] marking the beginning of her push to promote information about the seriousness of premenstrual syndrome. As we will see, one of her overriding concerns was the effect on women's performance at school and work and the cost to national economies of women's inability to work premenstrually.

Although Dalton's research fit in nicely with the postwar edging of women out of the paid work force, it was not until the mid to late 1970s that the most dramatic explosion of interest in PMS took place. This time there were no returning veterans to demand jobs for which women were suddenly "unqualified"; instead, women had made greater incursions into the paid work force for the first time without the aid of a major war.

> First single women, then wives, and then mothers of school-aged children were, in a sense, freed from social constraints against work outside the home. For each of these groups, wage labor was at one time controversial and debatable, but eventually employment became a socially acceptable—and even expected—act.[27]

Many factors were responsible for women's emergence in the paid work force: the second wave of feminism and stronger convictions about women's right to work, a lower birth rate, legislative support barring sex discrimination, increasing urbanization, and growth in educational opportunities for women.[28] It goes without saying that women's move toward center stage in the paid work force (as far away as equality still remains) is threatening to some women and men and has given rise to a variety of maneuvers designed to return women to their homes.[29] Laws has suggested that the recent burgeoning of emphasis on PMS is a "response to the second wave of feminism." I think this is a plausible suggestion, made even more convincing by the conjunction between periods of our recent history when women's participation in the labor force was seen as a threat, and, simultaneously, menstruation was seen as a liability.[30]

Turning to the premenstrual symptoms women themselves report, what views of the world of work are represented in their words? An overriding theme in the changes women articulate is a loss of ability to carry on activities involving mental or physical discipline. For example, from Lever's list (Table 1): "difficulty concentrating," "confusion," "forgetfulness," "lowered judgment," "lack of coordination," "decreased efficiency," "lowered school or work performance." Others report "an inability to string words together correctly"[31] or increased tendency to "fail examinations, absent themselves from work."[32] One book advises that if women find their ability to perform some aspects of work impaired premenstrually, they might organize their work so that "more routine work, for example, might be carried out during this time, and work requiring more concentration and care might be kept for other times when they feel more capable of it."[33] Competitive tennis players say their reaction times can be slower and professional singers say they lose voice control.[34]

It is no doubt completely understandable that in a society where most people work at jobs that require and reward discipline of mind and body, loss of discipline would be perceived negatively. Marx showed long ago that in a system in which an owner's profit is based on how much value can be squeezed out of laborers' work, the amount of time laborers would have to work and what they did (down to the precise movements of their hands and bodies) would be controlled by factory owners. Indeed, historically, when legislation forced a shortening of the working day, owners found it necessary to intensify labor during the hours remaining: "Machinery becomes in the hands of capital the objective means, systematically employed for squeezing out more labour in a given time. This is effected in two ways: by increasing the speed of the machinery, and by giving the workman more machinery to tent."[35]

Braverman and others have recently shown how scientific management, introduced in the late nineteenth century, has contributed to the deskilling and degradation of work: creative, innovative, planning aspects of the work process are separated from routine manual tasks, which are then extremely subject to finely tuned managerial control.[36] We are perhaps accustomed to the notion that assembly line factory work entails a bending of workers' bodies in time and space according to the demands of "productivity" and "efficiency," but we are less accustomed to realize that deskilling, leading to monotony, routine, and repetition, has increasingly affected not just clerical occupations and the enormous service industry[37] but the professions as well. Natalie Sokolov has shown that the lower edges of the medical and legal professions, edges into which women crowd, hardly have the independence, creativity, and opportunity for growth that we associate with professional work. Allied health workers, legal assistants, and lawyers who work in legal clinics may have to do tedious, boring work requiring minimum skills and little opportunity for advancement in knowledge or position.[38]

With respect to work, then, the vast majority of the population and all but a very few women are subjected to physical and mental discipline, one manifestation of what Foucault calls a "micro-physics of power," "small acts of cunning" in the total enterprise of producing "docile bodies."[39] What many women seem to report is that they are, during premenstrual days, less willing or able to tolerate such discipline.

An obvious next question is whether the incidence of PMS is higher among women subjected to greater work discipline.[40] One would also like to know whether there is any correlation between the experience of PMS symptoms (as well as the reporting of them) and factors such as class and race. Unfortunately, the PMS literature is nearly deaf to these kinds of questions. Although the incidence of PMS in relation to age, parity, and the existence of a male living partner has been examined (and is generally found to increase with each),[41] these are very crude indicators of the whole working and living environment of particular women. My own interviews, not directed toward women with PMS, turned up only four women out of 165 who described themselves as experiencing it, so I cannot speak to whatever patterns there may be.

Perhaps part of the reason a more sophisticated sociological analysis has not been done is that those who comment on and minister to these women do not see that women's mental and physical state gives them trouble, only because of the way work is organized in our industrialized society. Women are perceived as malfunctioning and their hormones out of balance rather than the organization of society and work perceived as in need of a transformation to demand less constant discipline and productivity.

Many PMS symptoms seem to focus on intolerance for the kind of work discipline required by late industrial societies. But what about women who find that they become clumsy? Surely this experience would be a liability in any kind of social setting. Perhaps so, and yet it is interesting that most complaints about clumsiness seem to focus on difficulty carrying out the mundane tasks of keeping house: "You may find you suddenly seem to drop things more often or bump into furniture around the house. Many women find they tend to burn themselves while cooking or cut themselves more frequently."[42] "It's almost funny. I'll be washing the dishes or putting them away and suddenly a glass will just jump out of my hands. I must break a glass every month. But that's when I know I'm entering my premenstrual phase."[43] Is there something about housework that makes it problematic if one's usual capacity for discipline relaxes?

On the one hand, for the numbers of women who work a double day (hold down a regular job in the paid work force and come home to do most of the cooking, cleaning, and child care), such juggling of diverse responsibilities can only come at the cost of supreme and unremitting effort. On the other hand, for the full-time homemaker, recent changes in the organization of housework must be taken into account. Despite the introduction of "labor-saving" machines, time required by the job has increased as a result of decline in the availability of servants, rise in the standards of household cleanliness, and elaboration of the enterprise of childrearing.[44]

To this increase in expectations placed on homemakers I would add the sense of how desirable it is to be "efficient" and "productive" at home, much as it is in the workplace. "Heloise's Hints" and similar columns in newspapers and magazines are full of tips on how to make every moment count, with clever ways of meeting perfectionist standards in the multitude of roles played by a homemaker. Perhaps the original idea came from one of the early masters of scientific management, Frank Gilbreth, who gave his name spelled backward (*therblig*) to the basic unit used in time and motion studies. Time and motion studies, carried out with increased sophistication today, are designed to break down the human actions involved in production into their component parts so they can be controlled by management.[45] It was a son and daughter of Gilbreth's who wrote *Cheaper by the Dozen,* a chronicle of how Gilbreth applied his notions of efficiency and productivity to his own household:

> Dad took moving pictures of us children washing dishes, so that he could figure out how we could reduce our motions and thus hurry through the task . . . Dad was always the efficiency expert. He buttoned his vest from the bottom up, instead of from the top down, because the bottom-to-top process took him only three seconds, while the top-to-bottom took seven.[46]

Perhaps the need for discipline in housework comes from a combination of the desire for efficiency and a sense of its endlessness, a sense described by Simone de Beauvoir as "like the torture of Sisyphus . . . with its endless repetition: the clean becomes soiled, the soiled is made clean, over and over, day after day. The housewife wears herself out marking time: she makes nothing, simply perpetuates the present."[47] Not only sociological studies[48] but also novels by women attest to this aspect of housework:

> First thing in the morning you started with the diapers. After you changed them, if enough had collected in the pail, you washed them. If they had ammonia which was causing diaper rash, you boiled them in a large kettle on top of the stove for half an hour. While the diapers were boiling, you fed the children, if you could stand preparing food on the same stove with urine-soaked diapers. After breakfast, you took the children for a walk along deserted streets, noting flowers, ladybugs, jet trails. Sometimes a motorcycle would go by, scaring the shit out of the children. Sometimes a dog followed you. After the walk, you went back to the house. There were many choices before nap time: making grocery lists; doing the wash; making the beds; crawling around on the floor with the children; weeding the garden; scraping last night's dinner off the pots and pans with steel wool; refinishing furniture; vacuuming; sewing buttons on; letting down hems; mending tears; hemming curtains. During naps, assuming you could get the children to sleep simultaneously (which was an art in itself), you could flip through *Family Circle* to find out what creative decorating you could do in the home, or what new meals you could spring on your husband.[49]

Here is Katharina Dalton's example of how a premenstrual woman reacts to this routine:

> Then quite suddenly you feel as if you can't cope anymore—everything seems too much trouble, the endless household chores, the everlasting planning of meals. For no apparent reason you rebel: "Why should I do everything?" you ask yourself defiantly. "I didn't have to do this before I was married. Why should I do it now?" . . . As on other mornings you get up and cook breakfast while your husband is in the bathroom. You climb wearily out of bed and trudge down the stairs, a vague feeling of resentment growing within you. The sound of cheerful whistling from upstairs only makes you feel a little more cross. Without any warning the toast starts to scorch and the sausages instead of happily sizzling in the pan start spitting and spluttering furiously. Aghast you rescue the toast which by this time is beyond resurrection and fit only for the trash. The sausages are charred relics of their former selves and you throw those out too. Your unsuspecting husband opens the kitchen door expecting to find his breakfast ready and waiting, only to see a smoky atmosphere and a thoroughly overwrought wife. You are so dismayed at him finding you in such chaos that you just burst helplessly into tears.[50]

Needless to say, by the terms of the medical model in which Dalton operates, the solution for this situation is to seek medical advice and obtain treatment (usually progesterone).[51] The content of the woman's remarks, the substance of what she objects to, escape notice.

A woman who drops things, cuts or burns herself or the food in this kind of environment has to adjust to an altogether different level of demand on her time and energy than—say—Beng women in the Ivory Coast. There, albeit menstrually instead of premenstrually, women specifically must not enter the forest and do the usual work of their days—farming, chopping wood, and carrying water. Instead, keeping to the village, they are free to indulge in things they usually have no time for, such as cooking a special dish made of palm nuts. This dish, highly prized for its taste, takes hours of slow tending and cooking and is normally eaten only by menstruating women and their close friends and kinswomen.[52] Whatever the differing demands on Beng as opposed to western women, Beng social convention requires a cyclic change in women's usual activities. Perhaps Beng women have fewer burned fingers.

For the most part, women quoted in the popular health literature do not treat the cyclic change they experience as legitimate enough to alter the structure of work time. However, several of the women I interviewed did have this thought. One woman expressed this as a wish, while reinterpreting what she had heard about menstrual huts (places of seclusion used by women in some societies when they are menstruating):

> [Does menstruation have any spiritual or religious significance for you?] I like the idea of menstrual huts a great deal. They intrigue me. My understanding is it's a mysterious thing in some ways. It infuriates me that we don't know more about it. Here are all these women—apparently when you get your period you go off to this hut and you hang around. [Because you're unclean?] That's what I feel is probably bull, that's the masculine interpretation of what's going on passed on generally by men to male anthropologists, whereas the women probably say, "Oh, yeah, we're unclean, we're unclean, see ya later." And then they race off to the menstrual hut and have a good time. (Meg O'Hara)

Another got right to the heart of the matter with simplicity:

> Some women have cramps so severe that their whole attitude changes; maybe they need time to themselves and maybe if people would understand that they need time off, not the whole time, maybe a couple of days. When I first come on I sleep in bed a lot. I don't feel like doing anything. Maybe if people could understand more. Women's bodies change. (Linda Matthews)

Still another woman foresaw some of the impediments to change, which might include the attitudes of other women:

> [What would you change if you could reorganize things to make life easier?] On that day I would just tell my TA's, "Well, listen, it's the first day of my period and that's why I wasn't in section and that's why I didn't hand in my paper." And that would be an acceptable thing, because really it can keep me from going to class, keep me from handing in a paper, even if I worked on the paper really long and hard. But it's interesting, a lot of times male TA's are much more understanding than female TA's. There are women who don't get it. So you're just like all this BS, you had your period. Guys, they don't really know what it is so they'll just take you on faith that it was really horrible and you couldn't deal with anything. (Anna Perdoni)

These women are carrying on what amounts to a twin resistance: to science and the way it is used in our society to reduce discontent to biological malfunction and to the integrity of separate spheres which are maintained to keep women in one while ruling them out of the other.

Given that periodic changes in activity in accord with the menstrual cycle are not built into the structure of work in our society, what does happen to women's work during their periods? Much recent research has attempted to discover whether women's actual performance declines premenstrually. The overwhelming impression one gets from reading the popular literature on the subject is that performance in almost every respect does decline. According to Dalton's influential account, women's grades drop, they are more likely to commit crimes and suicide, and they "cost British industry 3% of its total wage bill, which may be compared with 3% in Italy, 5% in Sweden and 8% in America."[53] Yet other accounts make powerful criticisms of the research on which these conclusions are based: they lack adequate controls, fail to report negative findings, and fail to report overall levels of women's performance in comparison to men's.[54] Still other studies find either increased performance or no difference in performance at all.[55]

Some women we interviewed expressed unforgettably the double message that women workers receive about PMS:

> Something I hear a lot that really amazes me is that women are discriminated against because they get their period. It makes them less capable to do certain kinds of work. It makes me angry. I never faced it in terms of my own personal experience, but it's something I've heard. I grew up thinking you shouldn't draw attention to your period, it makes you seem less capable than a man. I always tried to be kind of a martyr, and then all of a sudden recently I started hearing all this scientific information that shows that women really do have a cycle that affects their mood, and they really do get into bad moods when they have their periods. I don't know whether all of a sudden it gives legitimacy to start complaining that it's okay. I think I have a hard time figuring out what that's supposed to do. Then again you can look at that as being a really negative thing, medical proof that women are less reliable. It's proven now that they're going to have bad moods once a month and not be as productive. (Shelly Levinson)

I think the way out of this bind is to focus on women's experiential statements—that they function differently during certain days, in ways that make it harder for them to tolerate the discipline required by work in our society. We could then perhaps hear these statements not as warnings of the flaws inside women that need to be fixed but as insights into flaws in society that need to be addressed.

What we see from the list of PMS symptoms in Table 1 is not so much a list of traits that would be unfortunate in any circumstance but traits that happen to be unfortunate in our particular social and economic system, with the kind of work it requires. This consideration gives rise to the question of whether the decreases reported by women in their ability to concentrate or discipline their attention are accompanied by gains in complementary areas. Does loss of ability to concentrate mean a greater ability to free-associate? Loss of muscle control, a gain in ability to relax? Decreased efficiency, increased attention to a smaller number of tasks?

Here and there in the literature on PMS one can find hints of such increased abilities. Women report:

No real distress except melancholy which I actually enjoy. It's a quiet reflective time for me.

My skin breaks out around both ovulation and my period. My temper is short; I am near tears, I am depressed. One fantastic thing—I have just discovered that I write poetry just before my period is due. I feel very creative at that time.[56]

Others find they "dream more than usual, and may feel sexier than at other times of the cycle."[57]

A sculptor described her special abilities when she is premenstrual. "There is a quality to my work and to my visions which just isn't there the rest of the month. I look forward to being premenstrual for its effect on my creativity, although some of the other symptoms create strains with my family." Another woman, prone to depression, described in the journal she kept, "When I am premenstrual I can write with such clarity and depth that after I get my period I don't recognize that those were my thoughts or that I could have written anything so profound."[58]

I don't know what it is, but I'll wake up one morning with an urge to bake bread. I can hardly wait to get home from work and start mixing the flour, kneading the dough, smelling the yeast. It's almost sensual and very satisfying. Maybe it's the earth-mother in me coming out. I don't know. But I do enjoy my premenstrual time.[59]

I have heard that many women cry before their period. Well, I do too. Sometimes I'll cry at the drop of a hat, but it's a good crying. I'll be watching something tender on TV or my children will do something dear, and my eyes fill up. My heart is flooded with feelings of love for them or for my husband, for the world, for humanity, all the joy and all the suffering. Sometimes I could just cry and cry. But it strengthens me. It makes me feel a part of the earth, of the life-giving force.[60]

And from my interviews:

I dream very differently during my period; my dreams are very, very vivid and sometimes it seems that I hear voices and conversations. My dreams are very vivid and the colors are not brighter but bolder, like blues and reds and that's also very interesting. The last three days I feel more creative. Things seem a little more colorful, it's just that feeling of exhilaration during the last few days. I feel really great. (Alice Larrick)

I like being by myself, it gives me time to forget about what people are thinking. I like the time I don't have to worry about talking to anybody or being around anybody. It's nice to be by yourself. Time alone. (Kristin Lassiter)

Amid the losses on which most accounts of PMS focus, these women seem to be glimpsing increased capacities of other kinds. If these capacities are there, they are certainly not ones that would be given a chance to flourish or would even be an advantage in the ordinary dual workday of most women. Only the exception—a sculptor or writer—would be able to put these greater emotional and associative capacities to work in her regular environment. Perhaps it is the creative writing tasks present in most academic jobs that lead to the result researchers find puzzling: if premenstrual women cannot concentrate as well, then why are women academics' work performance and concentration better than usual during the premenstrual phase?[61] The answer may be that there are different kinds of concentration: some requiring discipline inimical to body and soul that women reject premenstrually and some allowing expression of the depth within oneself that women have greater access to premenstrually.

We can gain some insight into how women's premenstrual and menstrual capacities can be seen as powers, not liabilities, by looking at the ethnographic case of the Yurok.[62] Thomas Buckley has shown how the Yurok view of menstruation (lost in ethnographic accounts until his writing) held that

a menstruating woman should isolate herself because this is the time when she is at the height of her powers. Thus, the time should not be wasted in mundane tasks and social distractions, nor should one's concentration be broken by concerns with the opposite sex. Rather, all of one's energies should be applied in concentrated meditation on the nature of one's life, "to find out the purpose of your life," and toward the "accumulation" of spiritual energy.[63]

Michelle Harrison, with a sense of the appropriate setting for premenstrual women, says poignantly, "Women who are premenstrual often have a need for time alone, time to themselves, and yet few women actually have that time in their lives. One woman wrote,

'When I listen to music I feel better. If I can just be by myself and listen quietly, then the irritability disappears and I actually feel good. I never do it, though, or rarely so. I feel guilty for taking that time for myself, so I just go on being angry or depressed.'"[64] What might in the right context be released as powerful creativity or deep self-knowledge becomes, in the context of women's everyday lives in our societies, maladaptive discontent.

A common premenstrual feeling women describe is anger, and the way this anger is felt by women and described by the medical profession tells a lot about the niche women are expected to occupy in society. An ad in a local paper for psychotherapeutic support groups asks: "Do you have PMS?—Depression—irritability—panic attacks—food cravings—lethargy—dizziness—headache—backache—anger. How are other women coping with this syndrome? Learn new coping mechanisms; get support from others who are managing their lives."[65] Anger is listed as a symptom in a syndrome, or illness, that afflicts only women. In fuller accounts we find that the reason anger expressed by women is problematic in our society is that anger (and allied feelings such as irritability) makes it hard for a woman to carry out her expected role of maintaining harmonious relationships within the family.

> Serious problems can arise—a woman might become excessively irritable with her children (for which she may feel guilty afterwards), she may be unable to cope with her work, or she may spend days crying for no apparent reason. Life, in other words, becomes intolerable for a short while, both for the sufferer and for those people with whom she lives . . . PMT is often referred to as a potential disrupter of family life. Women suffering from premenstrual irritability often take it out on children, sometimes violently . . . Obviously an anxious and irritable mother is not likely to promote harmony within the family.[66]

This entire account is premised on the unexamined cultural assumption that it is primarily a woman's job to see that social relationships work smoothly in the family. Her own anger, however substantial the basis for it, must not be allowed to make life hard on those around her. If she has an anger she cannot control, she is considered hormonally unbalanced and should seek medical treatment for her malfunction. If she goes on subjecting her family to such feelings, disastrous consequences—construed as a woman's *fault* in the PMS literature—may follow. For example, "Doctor Dalton tells the story of a salesman whose commissions dropped severely once a month, putting a financial strain on the family and worrying him a great deal. Doctor Dalton charted his wife's menstrual cycle and found that she suffered from severe PMT. This affected her husband, who became anxious and distracted and so less efficient at his job. The drop in his commissions coincided with her premenstrual days. Doctor Dalton treated his wife and cured the salesman!"[67]

Not only can a man's failure at work be laid at the doorstep of a woman's PMS, so also can a man's violence. Although the PMS literature acknowledges that many battered women do nothing to provoke the violence they suffer from men, it is at times prone to suggest that women may themselves bring on battering if the man has a "short fuse": "[The woman's] own violent feelings and actions while suffering from PMT could supply the spark that causes him to blow up."[68] Or consider this account, in which the woman is truly seen as a mere spark to the man's blaze:

> One night she was screaming at him, pounding his chest with her fists, when in her hysteria she grabbed the collar of his shirt and ripped so hard that the buttons flew, pinging the toaster and the microwave oven. But before Susan could understand what she had done, she was knocked against the kitchen wall. Richard had smacked her across the face with the back of his hand. It was a forceful blow that cracked two teeth and dislocated her jaw. She had also bitten her tongue and blood was flowing from her mouth . . . [Richard took her to the emergency room that night and moved out the next morning.] He was afraid he might hit her again because *she was so uncontrollable* when she was in a rage. [Emphasis added.][69]

In this incident, who was most uncontrollable when in a rage—Richard or Susan? Without condoning Susan's actions, we must see that her violence was not likely to damage her husband bodily. A woman's fists usually do not do great harm when pounding a man's chest, and in this case they evidently did not. Ripping his clothes, however unfortunate, is not on the same scale as his inflicting multiple (some of them irreversible!) bodily injuries that required her to be treated in a hospital emergency room. The point is not that she was unable to injure him because of her (presumed) smaller size and lesser strength. After all she could have kicked him in the groin or stabbed him with a knife. The point is, she chose relatively symbolic means of expressing her anger and he did not. Yet in the PMS literature *she* is the one cited as uncontrollable, and responsible for his actions. The problems of men in these accounts are caused by outside circumstances and other people (women). The problems of women are caused by their own internal failure, a biological "malfunction." What is missing in these accounts is any consideration of why, in Anglo and American societies, women might feel extreme rage at a time when their usual emotional controls are reduced.

That their rage is extreme cannot be doubted. Many women in fact describe their premenstrual selves as being "possessed." One's self-image as a woman (and behind this the cultural construction of what it is to be a woman) simply does not allow a woman to recognize herself in the angry, loud, sometimes violent "creature" she becomes once a month.

I feel it is not me that is in possession of my body. My whole personality changes, making it very difficult for the people I live and work with. I've tried. Every month I say, "This month it's going to be different, I'm not going to let it get hold of me." But when it actually comes to it, something chemical happens to me. I can't control it, it just happens.[70]

Something seems to snap in my head. I go from a normal state of mind to anger, when I'm really nasty. Usually I'm very even tempered, but in these times it is as if someone else, not me, is doing all this, and it is very frightening.[71]

It is something that is wound up inside, you know, like a great spring. And as soon as anything triggers it off, I'm away. It is very frightening. Like being possessed, I suppose.[72]

I try so hard to be a good mother. But when I feel this way, it's as if there's a monster inside me that I can't control.[73]

I just get enraged and sometimes I would like to throw bookshelves through windows, barely feeling that I have control . . . these feelings of fury when there's nothing around that would make that necessary. Life is basically going on as before, but suddenly I'm furious about it. (Meg O'Hara)

I was verbally abusive toward my husband, but I would really thrash out at the kids. When I had these outbursts I tended to observe myself. I felt like a third party, looking at what I was doing. There was nothing I could do about it. I was not in control of my actions. It's like somebody else is taking over.[74]

Once a month for the last 25 years this wonderful woman (my wife) has turned into a .[75]

It is an anthropological commonplace that spirit possession in traditional societies can be a means for those who are subordinated by formal political institutions (often women) to express discontent and manipulate their superiors.[76] But in these societies it is clear that propitiation of the possessing spirit or accusation of the living person who is behind the affliction involves the women and their social groups in setting social relations right. In our own topsy-turvy version of these elements, women say they feel "possessed," but what the society sees behind their trouble is really their own malfunctioning *bodies*. Redress for women may mean attention focused on the symptoms but not on the social environment in which the "possession" arose. The anger was not really the woman's fault, but neither was it to be taken seriously. Indeed, one of women's common complaints is that men treat their moods casually:

> Sometimes if I am in a bad mood, my husband will not take me seriously if I am close to my period. He felt if it was "that time of the month" any complaints I had were only periodic. A few weeks ago I told him that until I am fifty-five he will have taken me seriously only half the time. After that he will blame it on menopause.[77]

Or if husbands cannot ignore moods, perhaps the moods, instead of whatever concrete circumstances from which they arise, can be treated.

> The husband of a woman who came for help described their problem as follows, "My wife is fine for two weeks out of the month. She's friendly and a good wife. The house is clean. Then she ovulates and suddenly she's not happy about her life. She wants a job. Then her period comes and she is all right again." He wanted her to be medicated so she would be a "good wife" all month.[78]

Marilyn Frye, in discussing the range of territory a woman's anger can claim, suggests: "So long as a woman is operating squarely within a realm which is generally recognized as a woman's realm, labeled as such by stereotypes of women and of certain activities, her anger will quite likely be tolerated, at least not thought crazy." And she adds, in a note that applies precisely to the anger of PMS, "If the woman insists persistently enough on her anger being taken seriously, she may begin to seem mad, for she will seem to have her values all mixed up and distorted."[79]

What are the sources of women's anger, so powerful that women think of it as a kind of possessing spirit? A common characteristic of premenstrual anger is that women often feel it has no immediate identifiable cause: "It never occurred to me or my husband that my totally unreasonable behavior toward my husband and family over the years could have been caused by anything but basic viciousness in me."[80]

Women often experience the depression or anger of premenstrual syndrome as quite different from the depression or anger of other life situations. As one woman described this difference: "Being angry when I know I'm right makes me feel good, but being angry when I know it's just me makes me feel sick inside."[81]

Anger experienced in this way (as a result solely of a woman's intrinsic badness) cannot help but lead to guilt. And it seems possible that the sources of this diffuse anger could well come from women's perception, however inarticulate, of their oppression in society—of their lower wage scales, lesser opportunities for advancement into high ranks, tacit omission from the language, coercion into roles inside the family and out that demand constant nurturance and self-denial, and many other ills. Adrienne Rich asks:

> What woman, in the solitary confinement[82] of a life at home enclosed with young children, or in the struggle to mother them while providing for them single-handedly, or in the conflict of weighing her own personhood against the dogma that says she is a mother, first, last, and always—what woman has not dreamed of "going over the edge," of simply letting go, relinquishing what is termed her sanity, so that she can be taken care of for once, or can simply find a way to take care of herself? The mothers: collecting their children at school; sitting in rows at the parent-teacher meeting; placating weary infants in supermarket carriages; straggling home to make dinner, do laundry, and tend children after a day at work; fighting to get decent care and livable schoolrooms for their children; waiting for child-support checks while the landlord threatens eviction . . .—the mothers, if we could look into their fantasies—their daydreams and imaginary experiences—we would see the embodiment of rage, of tragedy, of the over-charged energy of love, of inventive desperation, we would see the machinery of institutional violence wrenching at the experience of motherhood.[83]

Rich acknowledges the "embodiment of rage" in women's fantasies and daydreams. Perhaps premenstrually many women's fantasies become reality, as they experience their own violence wrenching at all of society's institutions, not just motherhood as in Rich's discussion.

Coming out of a tradition of psychoanalysis, Shuttle and Redgrove suggest that a woman's period may be "a moment of truth which will not sustain lies." Whereas during most of the month a woman may keep quiet about things that bother her, "maybe at the paramenstruum, the truth flares into her consciousness: this is an intolerable habit, she is discriminated against as a woman, she is forced to underachieve if she wants love, this examination question set by male teachers is unintelligently phrased, I will not be a punch-ball to my loved ones, this child must learn that I am not the supernatural never-failing source of maternal sympathy."[84] In this rare analysis, some of the systematic social causes of women's second-class status, instead of the usual biological causes, are being named and identified as possible sources of suppressed anger.

If *these* kinds of causes are at the root of the unnamed anger that seems to afflict women, and if they could be named and known, maybe a cleaner, more productive anger would arise from within women, tying them together as a common oppressed group instead of sending them individually to the doctor as patients to be fixed.

> And so her anger grew. It swept through her like a fire. She was more than shaken. She thought she was consumed. But she was illuminated with her rage; she was bright with fury. And though she still trembled, one day she saw she had survived this blaze. And after a time she came to see this anger-that-was-so-long-denied as a blessing.[85]

To see anger as a blessing instead of as an illness, it may be necessary for women to feel that their rage is legitimate. To feel that their rage is legitimate, it may be necessary for women to understand their structural position in society, and this in turn may entail consciousness of themselves as members of a group that is denied full membership in society simply on the basis of gender. Many have tried to describe under what conditions groups of oppressed people will become conscious of their oppressed condition. Gramsci wrote of a dual "contradictory consciousness, . . . one which is implicit in his [humans'] activity and which in reality unites him with all his fellow-workers in the practical transformation of the world; and one, superficially explicit or verbal, which he has inherited from the past and uncritically absorbed."[86] Perhaps the rage women express premenstrually could be seen as an example of consciousness implicit in activity, which in reality unites all women, a consciousness that is combined in a contradictory way with an explicit verbal consciousness, inherited from the past and constantly reinforced in the present, which denies women's rage its truth.

It is well known that the oppression resulting from racism and colonialism engenders a diffused and steady rage in the oppressed population.[87] Audre Lorde expresses this with power: "My response to racism is anger. That anger has eaten clefts into my living only when it remained unspoken, useless to anyone. It has also served me in classrooms without light or learning, where the work and history of Black women was less than a vapor. It has served me as fire in the ice zone of uncomprehending eyes of white women who see in my experience and the experience of my people only new reasons for fear or guilt." Alongside anger from the injustice of racism is anger from the injustice of sexism: "Every woman has a well-stocked arsenal of anger potentially useful against those oppressions, personal and institutional, which brought that anger into being."[88]

Can it be accidental that women describing their premenstrual moods often speak of rebelling, resisting, or even feeling "at war"?[89] It is important not to miss the imagery of rebellion and resistance, even when the women themselves excuse their feelings by saying the rebellion is "for no apparent reason"[90] or that the war is with their own bodies! ("Each month I wage a successful battle with my body. But I'm tired of going to war.")[91]

Elizabeth Fox-Genovese writes of the factors that lead women to accept their own oppression: "Women's unequal access to political life and economic participation provided firm foundations for the ideology of gender difference. The dominant representations of gender relations stressed the naturalness and legitimacy of male authority and minimized the role of coercion. Yet coercion, and frequently its violent manifestation, regularly encouraged women to accept their subordinate status."[92] Looking at what has been written about PMS is certainly one way of seeing the "naturalness" of male authority in our society, its invisibility and unexamined, unquestioned nature. Coercion in this context need not consist in the violence of rape or beating: sometimes women's violence is believed to trigger these acts, as we have seen, but other times it is the women who become violent. In a best-selling novel, a psychopathic killer's brutal murders are triggered by her premenstruum.[93] In either case, physical coercion consists in focusing on women's bodies as the locus of the operation of power and insisting that rage and rebellion, as well as physical pain, will be cured by the administration of drugs, many of which have known tranquilizing properties.[94]

Credence for the medical tactic of treating women's bodies with drugs comes, of course, out of the finding that premenstrual moods and discomfort are regular, predictable, and in accord with a woman's menstrual cycle. Therefore, it is supposed, they must be at least partially caused by the changing hormonal levels known to be a part of the cycle. The next step, according to the logic of scientific medicine, is to try to find a drug that alleviates the unpleasant aspects of premenstrual syndrome for the millions of women that suffer them.

Yet if this were to happen, if women's monthly cycle were to be smoothed out, so to speak, we would do well to at least notice what would have been lost. Men and women alike in our society are familiar with one cycle, dictated by a complex interaction of biological and psychological factors, that happens in accord with cycles in the natural world: we all need to sleep part of every solar revolution, and we all recognize the disastrous consequences of being unable to sleep as well as the rejuvenating results of being able to do so. We also recognize and behave in accord with the socially determined cycle of the week, constructed around the demands of work-discipline in industrial capitalism.[95] It has even been found that men structure their moods more strongly in accord with the week than women.[96] And absenteeism in accord with the weekly cycle (reaching as high as 10 percent at General Motors on Mondays and Fridays)[97] is a cause of dismay in American industry but does not lead anyone to think that workers need medication for this problem.

Gloria Steinem wonders sardonically

> what would happen if suddenly, magically, men could menstruate and women could not?
>
> Clearly, menstruation would become an enviable, boast-worthy, masculine event:
>
> Men would brag about how long and how much.
>
> Young boys would talk about it as the envied beginning of manhood. Gifts, religious ceremonies, family dinners, and stag parties would mark the day.
>
> To prevent monthly work loss among the powerful, Congress would fund a National Institute of Dysmenorrhea.[98]

Perhaps we might add to her list that if men menstruated, we would all be expected to alter our activities monthly as well as daily and weekly and enter a time and space organized to maximize the special powers released around the time of menstruation while minimizing the discomforts.

PMS adds another facet to the complex round of women's consciousness. Here we find some explicit challenge to the existing structure of work and time, based on women's own experience and awareness of capacities that are stifled by the way work is organized. Here we also find a kind of inchoate rage which women, because of the power of the argument that reduces this rage to biological malfunction, often do not allow to become wrath. In the whole history of PMS there are the makings of a debate whose questions have not been recognized for what they are: Are women, as in the terms of our cultural ideology, relegated by the functions of their bodies to home and family, except when, as second best, they struggle into wartime vacancies? Or are women, drawing on the different concepts of time and human capacities they experience, not only able to function in the world of work but able to mount a challenge that will transform it?

8 Birth, Resistance, Race, and Class

CESAREAN V
When they said We're gonna hafta
take the baby she started screaming
hysterically She thought they meant
actually take the baby away from her
(Ya know, put it up for adoption) So
they explained to her Oh no We don't
mean take the baby permanently We
mean take it temporarily Take it out of
your body but she wouldn't calm down
She just went on screaming
—Marion Cohen
Cesarean Poems

How does birth look through the eyes of women? If birth is currently thought of and described by medical texts as if it were work done by the uterus, and if women's bodies are consequently subjected to the same kinds of controls as workers in the workplace, do women, as workers often have, resist their condition?

Let us look first at some forms that workers' resistance has taken. David Noble in discussing industrial production describes how workers still controlled machine tools up until after World War II: they acted as skilled operators, directing a cutting or drilling tool to produce a part of a certain size and shape. As long as this was so, they were able to control the pace of their work: "to keep time for themselves, avoid exhaustion, exercise authority over their work, avoid killing gravy piece rate jobs, stretch work to avoid layoffs, exercise creativity and last but not least, to express their solidarity and hostility to management."[1] More dramatically, in the nineteenth and twentieth centuries, workers have waged struggles over the length of the workday so as to limit the amount of profit that management can extract from their labor. And of course workers have used the tactic of striking (refusing to work in an establishment under unjust conditions) since the early days of industrialization. In addition, from time to time, workers' resistance has focused directly on machines. The Luddites in early-nineteenth-century England vented their anger at the machinery of industrialization by destroying the machines so that management could not use mechanization to oppress the workers further.

Do women's efforts to resist procedures that they experience as intrusions on their autonomy resemble the efforts of workers? Childbirth activist literature can be seen (and sometimes describes itself) as guides to "self-defense in the hospital," and the methods women have developed are strikingly similar to those that workers have tried in the workplace.[2] First, consider the solidarity, power, and organization obtainable through forming groups. Women's health groups dedicated to resisting the way birth is handled in hospitals are many and well organized, from the Childbirth Education Association (CEA) and National Association of Parents and Professionals for Safe Alternatives in Childbirth (Napsac) to C/SEC and the Cesarean Prevention Movement.[3] Next, consider the workers' efforts to stall or slow down the rate of production. In birth, this takes the form of delaying the time at which a woman is first defined as being at each stage of labor. For example, the second stage of labor, from the time the cervix opens entirely until the baby is born, is defined by hospitals as lasting only a maximum of two hours. If the woman pushes in the second stage much longer than this, doctors usually intervene, pulling the baby out with forceps or performing a cesarean section. Doctors often make laboring women very acutely aware of the clock against which the labor must progress. One woman repeated what the doctors had told her, "We'll let her labor a couple more hours and then we'll do the cesarean" (Laura Cromwell). Midwives who attend home births have a variety of ways to buy time and "avoid starting the time clock." They do not count the second stage as starting until there is absolutely no trace of cervix visible or until the woman actually feels an urge to push (which may be delayed some time after full dilatation); or they may simply not examine the woman internally so that they do not know when she reaches full dilatation, instead letting the woman determine when she is ready for the next stage.[4]

Women themselves may try to reduce the amount of time they spend in the hospital by delaying admitting themselves as long as possible. Explicitly they often understand that this allows a shorter time for their labors to be defined as ineffective and the baby to be extracted operatively. Barbara Rothman has shown how the time allowed in the hospital for both first- and second-stage labor has been reduced steadily since the 1940s. In short, the uterus (as doctors see it) is being given less and less time to produce its product. One woman I interviewed was proud and satisfied at having delayed going to the hospital until she was 7 cm dilated and after that at not telling the resident she had had a previous cesarean section. (When he found out, total panic broke loose.) Another woman described herself as stalling continuously during her attempt to have a home birth, delaying by insisting an ambulance be called when the midwife decided she should go to the hospital, not divulging even to her husband the fact that the head was crowning when they reached the hospital: "a man in the elevator patted me on the head and said, 'Are you having a contraction, dear?' I thought, 'No, I'm having a baby!' Then my husband asked if I could feel it in my back, and I thought, 'Of course not, the baby's coming out!' How incredibly stupid people can be" (Sarah Lasch).

In one astounding and moving case, a woman whose first child had been born by cesarean managed to conceal what was happening from her husband, doctor, mother, and stepfather through four days and nights of labor in a two-bedroom cottage in the middle of a snowy winter.

> I made up my mind if I started labor I was going to keep it a secret. As much as possible because they were talking about time limits, you know, you start labor and you go fourteen hours and if you haven't had that baby yet, you just might get a c-section.
>
> So that night we went to bed and I didn't tell my husband too much because if he knew he'd say, "Awwww, let's go to the hospital, quick, it's forty-five minutes away." I made up my mind I'm not going to let anybody know until it's really there and I've got a head start on those doctors, and I'm not telling them until it's too late almost.
>
> [She tells how the doctor warned her not to wait until too late before coming in, as another patient had just done.] I'm like "Oh, no!!! Oh, no, unh, unh, I won't do that." I knew that woman was doing exactly what I was planning on doing. Get there seventeen hours too early and they've got you. They've got you hooked up and you feel like a guinea pig and your labor slows down and they're going, "Unh, unh, too bad, you didn't pass the test, you're not performing right, let's do a c-section." Down the hallway you go. I said that's not going to be me, but I didn't tell them that. But you know the pressure you're under, from different sides.

She managed to conceal what was happening through the breaking of her waters and transition, but when pushing started, she found it impossible to stop the loud sounds she wanted to make. Her husband took her to a nearby hospital rather than the one she was scheduled to attend, and there she was told she was only 1 cm dilated. By this time she had lost track of where she was in labor and felt a cesarean section was the only way out. "I said, 'I want a c-section. Here and now. Something is wrong with me. Besides being nuts, I am too small down here and the baby's hung up, she's up there and she can't get out.' I was having these grunting pains, I felt like I was grunting my guts out."

Thinking she was still in early labor, she and her husband set off for the hospital they were supposed to attend. By this time the insistence of the medical staff that she was still in early labor obliterated and confused what she was feeling. As she described it with hindsight:

> I just pictured my rectum stretching and opening up—what it was was my vagina. "If I could just get this out of me," I kept saying. "John, I think I've pooped." I felt a burning sensation, a tingling sensation as we got there, I felt something warm and twitching. I thought, "It's my pad and I've got this poop coming out, the size of a cannonball." [Arriving at labor and delivery] I said, "I'm pregnant and I need an enema, I'm okay, I just need an enema, then I want a c-section." So we go to the labor room door and they say, "There's the bathroom." I go and I sit and I feel down here, I felt this nice smooth little head. And I scream, "It's a baby!" These girls come in—nurses, whatever—grab me and throw me on the labor table. I got up and [makes guttural sound] she was out!
>
> Oh, I was so elated! "Leah! Look at you!!" We were just screaming with joy. Then the doctor walked in. After all those months of "Don't show up too late." "I promise, I promise," I did! He said, "Well, you escaped." I said, "Yeah." I just can't express the high, *the high,* the biggest I DID IT! I just wanted to scream. At that point those nurses didn't know I was a former c-section. (Janice Sanderson)

And then there is Luddism, workers' attacks on the machines themselves. Many women report simply unstrapping external fetal monitors the minute the nurse or doctor is out of the labor room. Others go for long walks around the hospital and do not return for hours or take a shower continuously so that monitors cannot be used.[5] And one woman described how she used physical force:

> I catch a glimpse of the flash of metal [the delivery room scissors for an episiotomy] going past me. We had done perineal massage every night [to make the perineum flexible enough to avoid tearing or the need for an episiotomy]. I grabbed the scissors; from the home birth I knew it took seven-

teen minutes to sterilize them. I wasn't even trying to take them out of the doctor's hand. It wasn't a logical thought at the time, I was just trying to break the sterile field. I said, "You aren't cutting me" and asked him to massage, but he didn't know how, he just went like this [gestures halfheartedly] and [after the baby emerged] said, "Well, you tore!" He wanted me to tear. (Sarah Lasch)

Finally, there is perhaps the most effective tactic of all, the equivalent of opening up your own shop or becoming your own boss: never going to the hospital at all and having your baby at home. This is as close as the birth movement gets to calling a strike on the industry of obstetrics: a more exact parallel to a strike (refusing to bear children at all) has not been necessary because, while factory owners *own* the means of production (machines, raw materials), leaving workers only their labor to withdraw, women have control over the means of reproduction (at least for the present, as we will see below) in the form of their own bodies. When they give birth at home, they own the whole shop and can be in charge of the entire enterprise.

The analogy between workers and women giving birth can be extended with a consideration of the consequences of replacing skilled workers with machines. As Noble shows, machines can seldom do everything that a human can. He describes how in machine tool industries, until very recently, management was forced to rely to some extent on the skill and knowledge of the machinist, to prevent machines under "automatic" control from destroying themselves or raw material.[6] But whenever human labor is replaced by machines, it is likely that some of the human skills on which the machines were modeled will become lost. This is partly a result of subdividing tasks that previously were performed by one person into smaller parts, each to be performed by different people or machines.

Examples of the "depletion of irreplaceable skills"[7] in birth are legion. Forceps themselves were introduced by male midwives in the seventeenth and eighteenth centuries and were part of what enabled these men to compete effectively against female midwives. Elizabeth Nihell, one eighteenth-century midwife, wrote that midwives who used forceps found them "at once insignificant and dangerous substitutes for their own hands, with which they were sure of conducting their operations more safely, more effectually, and with less pain to the patient."[8] Today midwives retain a substantial store of nontechnological knowledge that, they constantly remark, doctors do not know: how to use hands instead of forceps; how to assist in the delivery of a breech baby; how to avoid an episiotomy by having the mother push only *between* contractions, not during them, for the last few contractions before delivery.

Sometimes skills can be lost and then rediscovered. Dr. Brooks Ranney, who has practiced in Yankton, South Dakota, since 1949, reports on his development of "the gentle art of external cephalic version," or using one's hands to turn a baby that presents head up or sideways by pushing gently on the baby through the mother's abdomen. He notes the lack of training in the technique for young doctors, the importance of years of experience in his relearning the technique, and its success in preventing premature labors and cesarean sections associated with breech presentations.[9]

The desperation to which women can be reduced in efforts to retain control over their own bodies is captured in the story of one woman I interviewed who attempted to do an external version herself. She had planned a vaginal birth after cesarean at home attended by a midwife, but the labor was progressing in a way the midwife found alarming. The woman herself remained confident that everything was fine, thinking that since the baby was engaged in her pelvis in posterior position, face up rather than face down, her relatively long and difficult labor was to be expected. She told me that while her husband, friends, and midwife were waiting in the living room, talking about how it was imperative to go to the hospital, "I was in the other room doing an external version myself. I got her to the side; even my doctor, who is never shocked, just sat down [when I told him]. I didn't force her [the baby], I talked with her very careful in what I was doing. I really worked with her" (Sarah Lasch). An ambulance arrived, the baby turned all the way around on the ride to the hospital and was born a few minutes after arriving.

David Noble gives us the gripping story of how, through increasingly advanced technology, control of machine tools has been taken from workers, even to the point of achieving "the actual removal of the work force itself." Short of complete removal of workers, however, as workers have less and less control they lose the ability to determine the pace and process of work. In "continuous-control operations" such as oil refineries, strikes become useless because the company can continue operations at nearly full capacity even without the workers.[10] If doctors are like managers controlling the work that women's bodies do in birthing a baby, then will they stop short of actually removing the work force, the women themselves?

This question leads us to an examination of the rapidly developing reproductive technologies. Many have suggested that these technologies contribute to birth being seen as commodity production: eggs, sperm, wombs, embryos, and even babies are increasingly being bought and sold.[11] My analogy of birth and production shows that this analogy is a slight extension of existing tendencies: ever since labor in childbirth was defined as mechanical work done by the uterus, birth has been seen as the (re)production of goods.

What is changing is our ability to harvest components of the goods (embryo, eggs) earlier and to know in greater detail what the quality of the goods is.[12] The new prenatal technologies of amniocentesis and sonography are creating new norms for the standards of production, which, like the norms for progress in the stages of labor, are held up for women to conform to. My own second child was found by sonography to be too small for her gestational age (even though at present very little is known about the details of fetal growth). Doctors immediately raised the frightening specters of "microcephaly" or "placental insufficiency," which were only finally dissipated when she was born a normal seven pounds twelve ounces. My interviews with other women are full of anxieties, precautions, and interventions, from total bed rest to early induction of labor, that have come along with new standard expectations for fetal growth and development.

On the one hand, technology allows the development of new standards for fetal growth. On the other hand, doctors, husbands, and state governments are successfully using legal sanctions to force women to involuntarily alter their diets (stop taking drugs), alter their daily activity (be confined to a hospital for the last weeks of pregnancy), or undergo cesarean section to protect the rights of the fetus.[13] Although one could argue that at least some of these restrictions might benefit the fetus, they certainly give the woman no choice about sacrificing her rights to those of the fetus. The possibility exists that the woman, the "laborer," will increasingly drop out of sight as doctor-managers focus on "producing" perfect "products."

Even the simple ways women have used to resist technology's incursions—such as taking long walks in the hospital to avoid being hooked up to a fetal monitor—are becoming more and more difficult. One guidebook for a "happy and safe Cesarean childbirth experience" reports cheerfully: "Some hospitals are already using a new kind of monitor that uses radio signals instead of wires to transmit its reports about your uterine contractions and the baby's heartbeat, so you can be tracked wherever you are."[14] And even standard monitors that have to be strapped to the woman's belly or wired to the baby's head have the effect of removing the mother. The doctor or technician literally turns his back to the woman to see the screen or printout describing what is happening inside her. As one woman put it (her baby was registering fetal distress), "I can remember all these official-looking people coming in and hugging the monitor, when I needed the hug. All these smart people were hugging the monitor" (Laura Cromwell).

The laboring woman is being bypassed much as the laborer has been in continuous-process industries. But monitors sending radio signals pale in comparison with technology that can foster a wholesale abandonment of women's bodies: *in vitro* fertilization, together with "artificial wombs" growing babies "from sperm to term," a development described in 1973 as "just a matter of time,"[15] has been hailed as so important that it should be a national priority. The benefits of an artificial womb are said to be many, including the *reduction* of birth defects by "keeping the fetus in an absolutely safe and regular environment."[16] This seems to me the same kind of denigration of women's bodies that led to the belief in the 1950s—now largely discredited—that a scientifically formulated bottled milk product was better for babies than breast milk.

Yet it is not enough to look at the reproductive technology itself. As Noble cautions, the crucial factors in advancing technologies are the social relations of production which dictate who uses the new technology and for whose benefit.[17] My earlier discussion showed that, medically, birth is seen as the control of laborers (women) and their machines (their uteruses) by managers (doctors), often using other machines to help. Developments in the way doctors' machines are bought, maintained, and controlled may have an important impact on the entire process. As doctors become increasingly not owner-managers of small businesses but employees in corporations based in hospitals, they may themselves be losing control over their working conditions. Whereas the doctor previously was the manager who used machines to scientifically manage women's labor and the uterus was the machine the woman used to produce the commodity—the baby—now the corporation may be becoming the new manager who uses standards of performance (such as the number of patients treated per hour or the number of uses to which machines are put) to control the doctors' own labor.[18] The doctor may be becoming the new worker, who (with the help of machines) produces the baby.[19] This process would dovetail with some effects of the new technology that literally bypass the woman's body: using laparoscopy to extract an egg from the ovary or utilizing a glass dish for fertilization, for example. A newspaper photo of the first test-tube twins born in Baltimore shows them in the arms not of the mother or the father but of the attending obstetrician![20]

Beyond these indications, is there any sense in which the woman, like Noble's labor force, may disappear? Indirect evidence that some such change is occurring comes from differences in the way labor is described in successive versions of one of the major obstetrics textbooks, *Williams Obstetrics*. The tenth edition (1950) starts the section "Physiology and Conduct of Labor" with this sentence: "By labor is meant the series of processes by which the mature, or almost mature, products of conception are expelled from the mother's body."[21] In the fourteenth edition (1971), the passive verbal construction "are expelled" is kept, but the expelling agent is named as the mother: "Labor comprises the series of processes by which the mature, or nearly mature, products of conception are expelled by the mother."[22] Strikingly, in the sixteenth edition (1980), this sentence has been removed altogether! The section on the physiology of labor moves directly into consideration of the causes, stages, and (mechanical) forces involved in labor.[23] The woman appears only briefly in this chapter, in a paragraph on "intra-abdominal pressure."

It may be purely coincidental that the sentence describing labor as something done by the mother is missing from the 1980 *Williams*. Or, along with the trend toward remote-controlled monitors, and the attempt, with legal sanctions, to remove the mother from decisions about whether or not to undergo medical procedures designed to benefit the fetus, it may represent the next stage in the application of technology to birth: the actual removal of the laborer herself.

Even more startling, the latest (seventeenth) edition of *Williams* (1985) begins the chapter on the physiology of labor with two entirely new paragraphs before it discusses the causes of labor. The first paragraph begins:

> It is believed that the majority of our permanently institutionalized persons are those who are mentally or physically disabled as the consequence of an untimely birth. To suffer the agony of lifelong mental and physical impairment is surely the greatest tragedy that can beset a person, his or her family, society, and even the economies of the world. There can be no graver or more profound insult to the quality of life.[24]

Many would quarrel with these statements on their own terms. But when we view them in light of the increasing control exerted over the "relations of reproduction" by medical practice and the new technologies that support that control, we must wonder whether "quality of life" is really "quality control of life." When doctors ally with the "innocent and vulnerable" newborn *against* the sensibilities of its mother, surrounding an event that has great significance to her, and when certain knowledge about what causes the onset of labor as well as when it should ideally begin is not yet available, as the seventeenth edition of *Williams* acknowledges, one has to wonder whether the issue is not control over the woman and her birthing rather than protection of the innocent.[25] One has to ask: whose baby, whose life, whose birth, whose timing, and who has the power to decide?

In discussing birth so far, I have treated women as if they were a homogeneous group, more or less equally likely to be subject to a cesarean section depending on their individual health histories and more or less equally likely to find ways to resist. In truth, a woman's class background, together with her race, profoundly affects the kind of birth experience she will have in the hospital. At the most fundamental level, a woman's and her child's general health is linked to their socioeconomic standing. A study done in Baltimore showed that there was far greater mortality (death) and morbidity (illness) among the working class than among the corporate and upper middle classes and that the differential increased between 1960 and 1973.[26] Other studies have shown the greater incidence of chronic conditions among the working class nationwide.[27] What applies to adults applies also to children: statistics for white births in upstate New York showed "an inverse relation between neonatal, post-neonatal, and fetal mortality and socioeconomic level as measured by father's occupation, that is, mortality increased as the socioeconomic level decreased."[28]

When race is considered, the pattern repeats itself. In both black and white groups in New York City between 1961 and 1963, there was a 50 percent difference in neonatal mortality between children of professional and managerial fathers at the top and service workers and laborers at the bottom.[29] In addition, these figures reveal a dramatic difference between blacks and whites taken as a whole: the neonatal mortality rate among blacks in the highest socioeconomic class is close to the rate in the laborer and service worker category among the whites.[30] What applies to children applies to their mothers: between 1959 and 1961 the total maternal mortality rate for nonwhite women was four times the rate for white women, and the differential for all causes had increased over the previous ten years.[31]

These trends continue in the present:

> Afro-American men and women experience greater morbidity and mortality from certain cancers, and from hypertension, diabetes, and other occupational and chronic diseases. For example, if we look at diseases of the heart, the first leading cause of death, statistics for 1977 show an overall age-adjusted heart disease rate of 322 per 100,000 for Afro-American men, compared with a rate of 294 per 100,000 for Euro-American men; for Afro-American women, we find an age-adjusted rate of 204 per 100,000, compared with 137 for Euro-American women. For cancer, the second leading cause of death, Afro-American men had an age-adjusted death rate from cancer of 222 per 100,000 in 1977 compared with a rate of 133 per 100,000 for Euro-American men; Afro-American women had a rate of 130 deaths per 100,000 compared with a rate of 108 per 100,000 for Euro-American women.[32]

The causes of these differentials surely lie within the different social circumstances of different groups. The effects of poverty on health through inadequate diets, substandard housing, and inability to afford medical care or insurance are well known.[33] The less obvious effects of occupational hazards on health are beginning to be explored: since workers in the United States are incorporated into the labor market along racial, ethnic, and class lines, work hazards—chemicals that cause cancer or other debility, lack of benefits such as sick pay or insurance, and long hours—fall disproportionately on those at the bottom of the occupational ladder.[34]

In addition, the effects of stress produced by the way work is organized in our society figures in health concerns: the relationship between unemployment and chronic disease[35] and the relationship between stress produced by jobs with low decision control and high psychological demand—jobs such as mail worker, garment stitcher, cashier, sales clerk, waiter, or telephone operator—and cardiovascular illness. Despite the common stereotype of high-powered executives who have heart attacks (and many do, of course) these individuals are not in the highest risk group for heart disease because alongside the high demands they experience from their jobs, they enjoy a great deal of control over their time and activities. Instead, workers in jobs such as garment stitcher, sales clerk, or waiter fall into the highest risk group because their jobs combine high demands and a low degree of control over their time and activities.[36]

When we look specifically at the types of births women of different classes and races experience, in particular at cesarean sections, two different expectations emerge. If cesarean sections are an expensive and scarce resource, more available to those who have the money to pay and the confidence and status to insist they need them, then we would expect to see *fewer* cesareans down the scale of class and race. Those who already have usually get more. The more cesarean sections become a resource that benefits hospitals and doctors, the more this pattern would be exacerbated. If the cesarean becomes a source of higher physicians' fees and hospital charges than vaginal delivery, we would expect to find more cesarean sections performed on women at the top of the class and race scales. Alternatively, or perhaps superimposed on this pattern might be another pattern, in which, along the lines of my analysis in Chapter 4, cesareans represent a means of control of women and their births often exerted when they are not in a position to resist. If this were so, we would expect the opposite: *more* cesareans down the scale of class and race.

Any such scrutiny will be complicated by the additional factor of need. Do poorer women have more difficulty birthing because of poorer nutrition or more stressful living conditions? Or, conversely, do poorer women have smaller babies because of stress and poor nutrition, and thus *less* difficult labors?

The statistical picture is confusing, to say the least, and it raises as many questions as it answers. But perhaps trying to ask questions of the existing data will inspire others to produce the information that could clear up these puzzles. A study done on New York City births from 1968 to 1977 showed what we would expect according to the first pattern outlined above: cesarean sections appear to be a scarce resource allocated more frequently to those with higher education and with the money to pay private doctors.[37] More recently, the dominant pattern is clearly that women in more privileged socioeconomic classes (who give birth in proprietary or voluntary hospitals with the aid of private doctors) get substantially more cesareans.[38]

But we must not be too quick to assume that the greater ability of some women to pay for cesarean sections will ensure that they always get more. In the New York City study, although overall there were minor differences among racial groups, in both 1968–69 and 1976–77 Puerto Rican whites had a higher cesarean section rate than other whites, and the differential increased between the two time periods.[39] Whatever the explanation of this, it certainly runs counter to the overall pattern. One possible explanation of these statistics could be based on a historical tendency that Hurst and Summey noticed in their research:

An historical pattern prevails: technology is introduced on poorer patients where it is tested, and where physicians learn to use the new methods, devices or medications; if accepted, it is then passed on to the private sector and becomes the preferred "modern" style of practice. Once the "testing" period has passed, and the new technology with its attendant protocols becomes part of the regular training experience, general service intervention rates resulting from the new technology tend to level off or even drop.[40]

Other factors that might contribute to a disparity in rates between different segments of the population can only be hinted at. A study done in upstate New York showed few differences between whites and blacks in 1951–52 and 1960–62. But broken down according to the type of complication that led to a cesarean section, blacks got far more sections for the diagnosis "dyscotia" than whites.[41] This is an indication to me of the second pattern outlined above—cesarean section as a means of control—because dyscotia is a vague, catchall diagnosis including all forms of uterine inertia (insufficiently strong or coordinated contractions), as well as feto-pelvic disproportion (insufficient room for the baby to pass through the pelvis).[42] This diagnosis arises directly out of the imposition of time limits on the rates of reproduction discussed in Chapter 4 and is centrally involved in exerting control over women's labor. Put another way, when there are clear clinical indications of fetal or maternal danger (bleeding, high blood pressure, prolapsed cord) more white women get a cesarean section, but when the labor is long or the rate of progression is slow, more black women get them. Undoubtedly some of these cases are truly life-saving and necessary, but one has to wonder why this difference runs along racial and class lines.

Similarly, Lillian Gibbons reports an increase in the diagnosis of dyscotia among blacks in Baltimore between 1968 and 1973, especially among teenagers, and raises questions about medical treatment that might be contributing to this increase. Perhaps, she speculates, the labors of black women are more often induced by pitocin, a procedure that often leads to the diagnosis "failure to progress" and the need for a cesarean.[43] The causal links between pitocin and "failure to progress" can be either that the artificially begun or speeded-up labor does not sustain itself or that pitocin produces such powerful and painful contractions that the woman cannot tolerate them, leading to early need for medication, which itself contributes to "uterine inertia."

In addition, both the New York City study and the Baltimore study show a puzzling difference in the birth weights of babies of black and

Table 3 PERCENTAGE OF LIVE BIRTHS DELIVERED BY CESAREAN SECTION BY RACE OF CHILD, MARYLAND, 1970–83

	Race of child		
Year	*Total*	*White*	*Nonwhite*
1970	5.7	5.2	7.4
1971	6.6	6.1	8.1
1972	8.2	7.5	10.3
1973	9.4	8.6	11.4
1974	11.1	10.3	13.5
1975	12.8	12.2	14.5
1976	15.1	14.6	16.3
1977	16.6	16.1	17.9
1978	18.3	17.6	19.8
1979	19.9	19.2	21.3
1980	20.6	19.7	22.3
1981	21.5	21.0	22.7
1982	22.5	22.0	23.4
1983	23.6	23.3	24.1

Source: Maryland Center for Health Statistics, 1984

white women who get cesarean sections. Those black women who have small babies have an appreciably lower rate of cesarean section.[44] One has to wonder what this means, especially given the association between low-birth-weight black babies and a higher incidence of infant mortality.[45] Are these black babies not registering fetal distress? If so, is it because their mothers are not hooked up as often to fetal monitors? Or is their distress being interpreted differently than the distress of white babies? Many puzzles, no answers.

The Baltimore study only adds to the confusion. In both 1968 and 1973, "adjusted cesarean section rates were uniformly higher for blacks." In some hospitals the rates for blacks were one and a half times higher than for whites.[46] In the state of Maryland as a whole, whites have been receiving fewer cesareans than non-whites since 1970. (See Table 3.) Any thorough exploration of the significance of this difference would have to take into account the age composition of the population, parity, and many other factors. It may simply be that black mothers, coming disproportionately from lower socioeconomic groups, are at greater medical risk, more often need cesarean sections, and get more of them in proportion to their need. But if we allow the possibility that cesarean rates are higher than would benefit mothers and children, the possibility arises that those less in a position to resist interventions (inductions of labor and cesarean sections) receive more of them.[47]

An additional factor in the medical care given minority women is, of course, the attitudes inculcated in doctors in medical school and afterward. Dan Segal has discussed the ways in which racism is explicitly used in medical school culture as a way of further entrenching the superior standing of doctors over the rest of the population.[48] And these attitudes can affect a doctor's practice. During her fieldwork in hospitals, Diana Scully found that poor black patients are openly regarded as objects of scorn. As doctors explained it to her, "It's much nicer to have a nice gravida two or three [a woman who has had two or three pregnancies] that can verbalize all her problems to you rather than a gravida fourteen who can't even remember all their names let alone how much she weighed or where they were born. But that is my own upbringing." Or "I'm sure you have seen what we have here. We have a lot of young girls having their first baby, some fourteen, even twelve; they are scared to death. They scream and shout, they won't lie in bed, they pull out their IV's, they won't let you examine them, that kind of business." Some of the practical consequences of these attitudes are that doctors are reluctant to explain things to such patients. They may be verbally harsh, strap the women down, tell them they are not in pain, refuse to allow husbands in the delivery room, refuse comfort, or refuse pain medication.[49]

These doctors are probably not very different from many whites in the general population in their disparagement of the lives and behavior of black women. From time to time in my interviews, white women voiced the same views. One white woman described her roommate in the hospital, a black 15-year-old having her first baby, and the girl's mother, pregnant with her thirteenth child. "The mother looked at me strange, wanted to know what I was doing and I said I was nursing the baby, and I said, 'Haven't you nursed any of your babies?' She laughed, she was so embarrassed. She wasn't embarrassed at having a grandchild from her daughter who was fifteen and she was pregnant with her thirteenth child, but *that* [nursing] would have been real embarrassing" (Vivian Randals). This woman's disparagement of her roommate's family was unfortunate but far less likely to directly affect the roommate than her doctor's disparagement because of the amount of control he had over her while she was in the hospital.

Minority women's accounts of birth reflect what it is like to be on the receiving end of medical disapproval:

They put me out, a mask over my face and I was out. When I woke up I think it was the next morning. [What was the reason?] They said that since I was in labor for fourteen hours and I just got up to like a 6 [cm dilated] and he had turned, they just rushed me to the delivery room and they put me out. It was so fast; everybody was talking at the same time. I ain't know what was going on. So they put me out and I didn't ever know that I had a baby or nothing. The next morning when I got up and somebody was waking me up, I wasn't informed at all. When I questioned it, my doctor said, "Well your pelvic area was too small for the baby to pass through." He turned and that was that and I never questioned it anymore. (Linda Matthews)

[What would you like the role of the doctors and nurses to be?] Some of those doctors and nurses, they are all smart and they know you're in pain and they say stuff like "You ain't say that when you got it" [You didn't complain when you had sex and got pregnant.] See, if they tell me that, I be mad, I'm all in pain, I don't want to hear all that. I want them to cooperate with me. (Juliet Latham)

My main problem when I was in labor was that I had a hard time relaxing myself. And then they kept saying that they were going to give me some type of medication to help relax me if I don't relax myself. And so, that's what they ended up doing right before they took me to the delivery room, because I was knocking down everything, knocking down the IV pole, everything! [How did you feel about getting the medication?] At first I said, "No, don't give it to me," because I said, "I'll just give me a couple of minutes, I'll see if I can try to relax." And then they said, "No, you've been trying to do it for the last couple of hours or so and you haven't been able to relax. So all you got to do is just lay there and relax." (Janice Peterson)

Not all the black women who had these kinds of experiences were working class, but it is possible that medical personnel's prejudice is as much based on their assumptions about class as about race. Although my numbers are too small to provide a definitive answer, it is striking that nearly half of the twenty-eight black working-class women we interviewed about recent birth experiences related incidents of this kind, while none of the white working-class women did. One white woman described behavior that doctors no doubt would have found objectionable in black women but that they tolerated in white women:

This one nurse, she is talking all nice to me and everything and I am smacking her. She walked out and my son's father walked out with her and I remember asking, "How am I doing?" And the nurse said, "She is doing fine." All she was trying to do was rub my back but I didn't want to be bothered. And my boyfriend just said, "She is doing fine. She is in there beating the crap out of that nurse." I apologized to her much later. And she said, "I can understand." Even my doctor got bit. He was going to another room with another [white] lady and she just went ahead and bit him. He wasn't too pleased either. But he went on helping her. (Patricia Barton)

Whether the dominant mechanism in the differing treatment of women in labor is race or class, it is evident that both profoundly affect birthing in our society. The ways women are able to resist what they dislike about the medical treatment of birth is clearly affected by their class and their race. Young black women in a very real sense have more to resist: not only a greater chance of having interventions and operations used on them, but the demeaning burden of racism instantiated in the ways they are treated. For a white middle-class woman, the salient issue may be to stall going to the hospital so the clock cannot be started or to organize and demand that all hospitals in the region install birthing rooms; for a white working-class woman, stalling may be an issue, but behind it lurks the larger issue of finding a way to pay for prenatal, obstetrical, or infant care; for a black working-class woman, the issues of stalling and paying may be crucial, but even if she contends with them, she still may have to find a way to avoid downright mistreatment or to manage to have matters explained to her at all.

9 The Creation of New Birth Imagery

The tradition of the dead generations weighs like a nightmare on the brain of the living.
—Karl Marx
The Eighteenth Brumaire of Louis Bonaparte

In the kind of resistance I have described so far, women are held to the terms of a debate whose issues are set by the medical profession. The issues to be discussed, the actions to be considered, are determined by doctors and hospitals, not by birthing women. The movement to bring consumer pressure on hospitals and doctors shows this very clearly. For example, the "Pregnant Patient's Bill of Rights" specifies that a pregnant patient has the right to be informed about the risks to her safety or her baby's from any drug, procedure, or therapy. This movement does not, nor could it reasonably, reject entirely the terms of the medical model of birth.[1] Another example is the contradictory concept of "family-centered cesarean." This is the notion (which undoubtedly improves the c-section experience) of giving parents some choice about the type of anesthesia, allowing partners to be present, letting babies and parents be together immediately afterward, and providing supportive care for the mother.[2] Laudable as these efforts are when a cesarean is the only possible safe birth method, one has to wonder whether it is not a relatively easy way for doctors and hospitals to make surgical birth more pleasant; far easier, for example, than actively encouraging most previous cesarean women to have vaginal birth or actively encouraging most healthy women to give birth at home.

Evaluated as a way to capture the significance of birthing, the analogy between production of goods and production of babies is profoundly misleading. If consumer efforts to resist are necessarily held within medical views of birth, including the view that birth is a form of production, then how do we attain a genuinely different vision? What would women, if allowed to arrange things to fit their proclivities and desires from the ground up, do at birth? What imagery would they bring to the experience, what practical arrangements would they seek, what technology would they develop, and under what conditions would they use it? These questions can be answered only by women—of different ethnic and class backgrounds—gathered together in groups reflecting on and acting on their own experiences. This is where we must also look for an answer to the question What would be lost if women were removed as laborers from the process of birth? There is a compelling need for new key metaphors, core symbols of birth that capture what we do not want to lose about birth.

Nancy Cohen and Lois Estner's "purebirth" is one attempt to develop such a new metaphor:

> Birth that is completely free of medical intervention. It is self-determined, self-assured, and self-sufficient, without necessarily being solitary . . . purebirth has no stages. Rather, it is a continuation of a creative energy that began with conception and will grow through years of nurturing.[3]

The concrete metaphors they choose to explain their concept stress a process that develops from within and the continuity of this process with the past and future. *A river:* "the continuous flow of labor narrows into an intense stream of life-filled birth"; *a ripening fruit:* "like fruit ripening on a tree, birth takes time. If we start too soon or try to rush, it will be like picking unripe fruit: harder work, longer hours, and possible damage to the crop."[4]

Other writers focus more explicitly on the energy people feel to be present during birth: "At birth, you, the laboring woman, are the channel for the Life Force, the energy of creation and transformation."[5] They choose metaphors that capture how the woman seems to "ride" this energy, actively adapting to it: "It is not possible to control your labor, but it is both possible and necessary to control yourself, in the way that a surfer controls himself in riding the big waves by maintaining equilibrium at the same time that he surrenders to them." Dwelling on heightened energy leads these writers to stress the similarities between the experience of birthing and the experience of sexual union: "Birth is fundamentally a creative act, as is the act of sexual union. The quality and intensity of the energy present and the ultimate surrender during both events are closely related."[6]

Out of a similar concern with energy, residents of the Farm, a spiritual community in Tennessee, have created completely new metaphors for conceptualizing "contractions." "During labor your uterine muscles contract at intervals and finally push out the baby. While this is happening, your cervix is thinning and opening. We call these regular bursts of energy 'rushes.' Labor progresses best if you pay attention to the expansion rather than to the contraction."[7] A mother of three at the Farm describes her experience: "Contractions don't have to hurt. They are energy rushes that enable you to open up your thing so the baby can come out. If you have the attitude that they hurt, then you'll tense up and not be able to completely relax and it will take the baby longer to come through and you won't have any fun either. It is a miracle to be able to create more life force and there is no room for complaining."[8]

While the view of birth as concentrated life force allows the mother to be either a passive vessel through which the force flows or an active participant "riding" the energy, other views place the woman in an unambiguously active role. Consider the notion of "positive birthing," which focuses on functional integration of all of a woman's parts—her memories of the past, hopes for the future, her mind and body. The central image is the dance: "Positive birthing is a psychophysiological (mind-body) approach to childbirth that focuses on the harmonious integration of body, mind, heart, and soul. The mind and body, when aligned, dance in rhythm and unity, surrendering willingly to the new life coming forth."[9]

Another view based on another key metaphor also places the woman in an active role. Gayle Peterson says: "Birth is a journey . . . The view of pregnancy and birth as a journey inward has begun at the end of the first trimester. Birth becomes an opportunity for psychological growth and an event to which a laboring woman relates intimately and uniquely, weaving a learning experience all her own."[10]

Many women search for metaphors that capture the sense of acting and doing: running a marathon, climbing a mountain, swimming an Olympic race,[11] skiing down a mountain (Sally Xenos); "moving a grand piano across the room: that hard, but that satisfying, to feel it moving along."[12]

Key metaphors can buttress existing organizations of experience and practice or show the way to new ones. If birthing is just like production in our society, then we might as well let it be managed, kept efficient, and eventually be done entirely by machines. But if birthing is part of our river of life, part of an inward and outward journey, we would want it to occur where we live our lives, at home, surrounded by friends and family. If birthing is hard work, but work that is rewarding in the doing of it, then we would want to be given time, encouragement, and support to do it ourselves. If birthing is a profound, heightened experience involving deep (often ecstatic) feelings and perception of powerful forces in the world, we would *want* the experience, period.

An example of how a key metaphor can organize experience and practice in very specific ways comes from a group centered in Kansas devoted to "do-it-yourself" home birth. Using references to Christian scriptures, this group asserts one metaphor at the expense of all others: birth is an "intimate husband/wife love encounter."[13] When genital sex is chosen as the key metaphor for what birth is, a number of consequences for action follow. Husband and wife will give birth alone, in private, just as they would when engaged in other sexual behavior.[14] To help the process along, just as in other forms of sexuality, lovemaking, fantasizing, hugging, kissing, caressing are the most relevant means.[15] The need to be private dictates that no birth attendant be present, but "when a midwife physically assists in the birth of a baby it is a homosexual experience . . . it is not what the midwife says that makes it a lesbian experience. It is what she does. She could be a deaf mute. But if she is handling the private part of a woman who is having a genital expression (and that is what birth is), then that is a homosexual experience. Only the husband of the birthing woman should be touching her there."[16]

Different lives, different births. But in spite of the diversity, a common theme runs through those accounts of birth in which women are self-consciously trying to create meaning in the event. That theme—finding a sense of wholeness—is a response to a call that often comes from the women's health movement: "to reintegrate the whole person from the jigsaw of parts created by modern scientific medicine."[17] The call is clear, and so is the accompanying warning that reintegration cannot be done simply, for "the parts biomedicine currently recognizes cannot be reassembled into a whole."[18]

We should heed this warning, for it would be very surprising if advocates of new birth imagery managed to create visions that were completely without contradictions, completely free of the dominant ideology of our society. An example of an effort to transform the experience of birth that is fraught with contradictions is occurring at Michel Odent's clinic at Pithiviers, France. Although the clinic is in France, Odent's book, *Birth Reborn,* has been championed by birth activists in the United States. Not a few of them have visited Pithiviers to see for themselves the kind of births Odent describes:

> Labor is allowed to progress without recourse to amniotomy nor to any drugs for induction or augmentation of contractions. Neither analgesics nor anesthesia is used except for operative deliveries. [The rate of cesarean section is kept very low, at about 5%.] When labor has progressed to the point where the mother and father feel more comfortable in one of the birth rooms, they walk there. The mother chooses any position she wishes, changing position at will throughout labor and delivery. Most women prefer to walk, kneel or sit during contractions, or they may take warm baths or relax in one of two plastic swimming pools filled with warm water. Every effort is made to help women find comfortable positions, and especially to allow their instinctive or viscero-affective brains to dictate their behavior. Delivery occurs in whatever place or position the mother finds comfortable. Often, this position is a supported standing-squat, in which the mother is supported under the arms and shoulders while she flexes her knees slightly and bears down.[19]

By all accounts, what Odent has done for women's birthing experience is extraordinary. One has only to compare the visual iconography of birth in a hospital (Figure 25) to the strong, active, upright vision of women at Pithiviers (Figures 26 and 27) to see the dramatic difference. But the cost is high. Odent finds it impossible to provide women this new setting for birth without describing what happens to them in it as a return to an animal-like, childlike state.[20] Although he says his intent is to let *women* give birth to their children,[21] he interprets what they *do* like this:

> Women seemed to forget themselves and what was going on around them during the course of an unmedicated labor . . . They get a faraway look in their eyes, forget social conventions, lose self-consciousness and self-control. Many let out a characteristic cry at the moment of delivery . . . I have found it very difficult to describe this shift to a deeper level of consciousness during labor. I had thought of calling it "regression," but I know that the word sounds pejorative, evoking a return to some animal state. "Instinct" is a better term, although it, too, resonates with moralistic overtones . . . There is nothing shameful or sexist in recognizing that instinct plays a part in our behaviors, especially those that exist at the intersection of nature

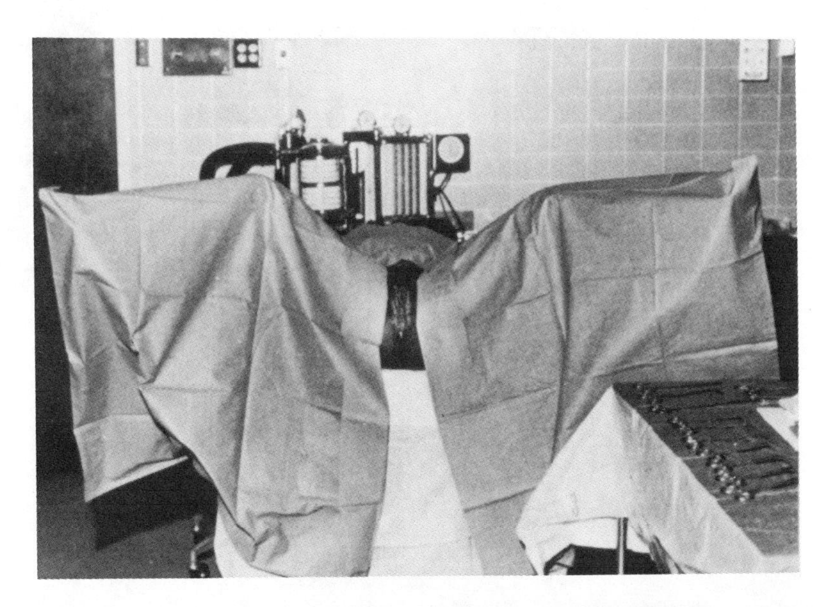

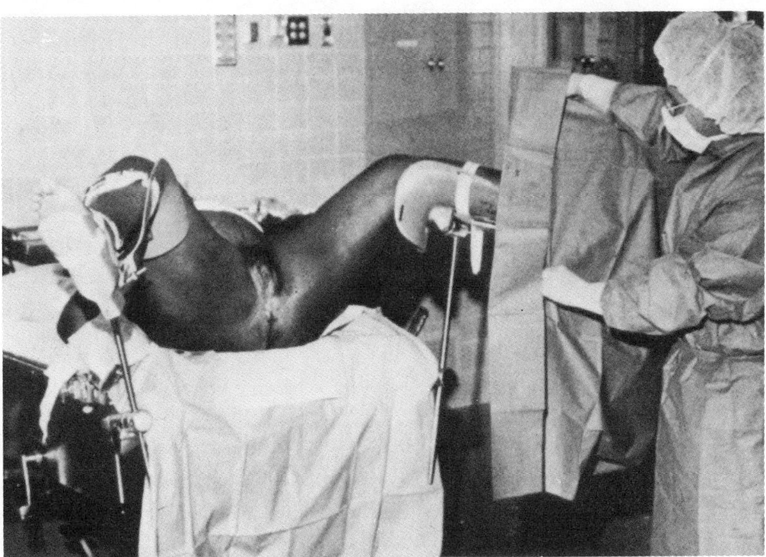

Fig. 25 (A) The standard position for delivery used in most hospitals in the United States. The woman's legs are held up and apart in stirrups while she lies flat on her back. (B) The vulva, perineum, and adjacent areas have been thoroughly scrubbed in preparation for the birth. Sterile drapes entirely cover the woman's abdomen. (Pritchard and MacDonald 1980:416, figs. 17-4A and 17-4B. Reprinted by permission of Appleton-Century-Crofts.)

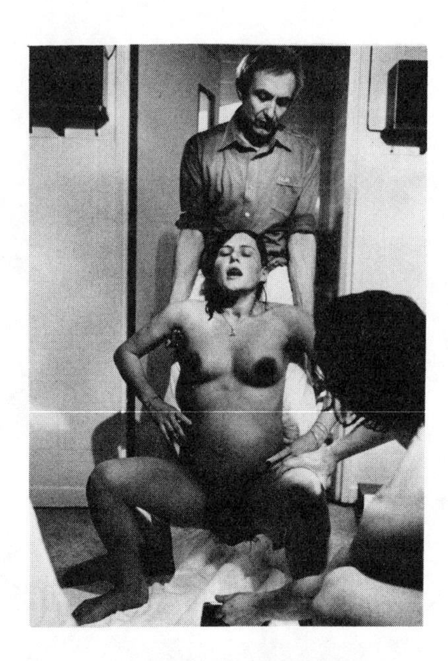

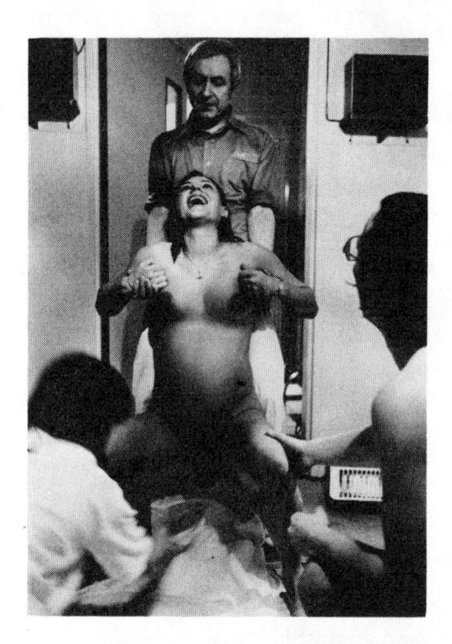

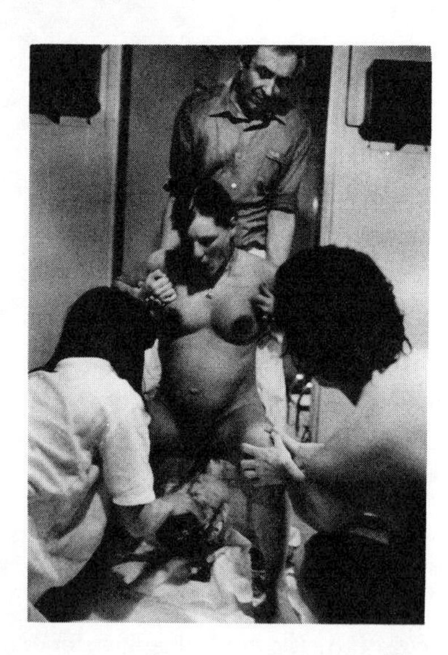

Fig. 26 Birthing at Michel Odent's clinic in Pithiviers, France. Women find it comfortable to squat while being supported under the arms from behind as they push their babies out. (From *Birth Reborn* by Michel Odent, translated by Jane Pincus and Juliette Levin, p. 48. Copyright © 1984 by Michel Odent. Reprinted by permission of Pantheon Books, a division of Random House, Inc.)

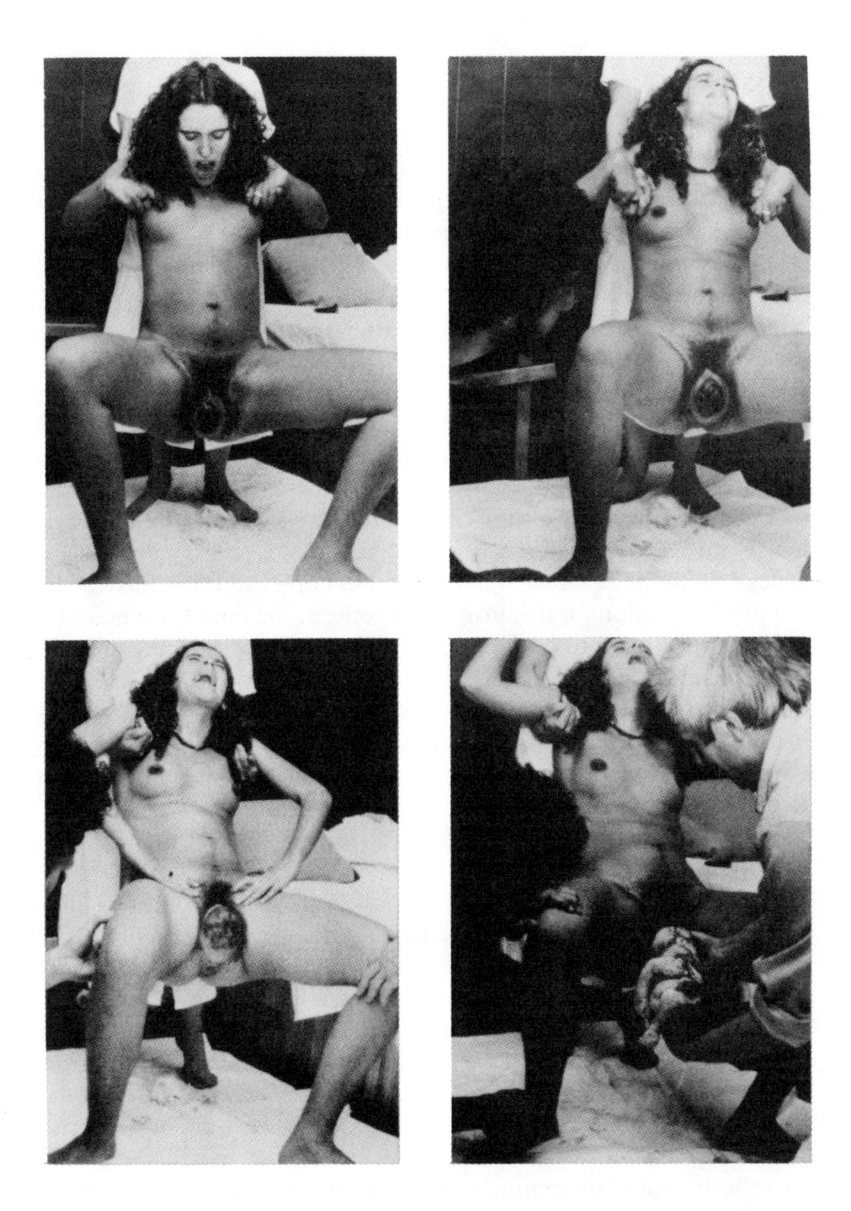

Fig. 27 Birthing at Michel Odent's clinic in Pithiviers, France. (From *Birth Reborn* by Michel Odent, translated by Jane Pincus and Juliette Levin, p. 57. Copyright © 1984 by Michel Odent. Reprinted by permission of Pantheon Books, a division of Random House, Inc.)

and culture, such as lovemaking, labor, or the newborn's search for the mother's nipple.[22]

What has happened? In Odent's view, birthing women are perceived as moving back in time and down the evolutionary tree to a simpler, animal-like, unselfconscious state.[23] This assessment must be viewed in light of the historical exclusion of women from "culture"—that higher activity of men—and the exclusion of women's culture (such as their writings) from the mainstream.[24] It is ironic that Odent's efforts to give birthing back to women occur at the cost of reasserting a view of women as animal-like, part of nature, not of culture. Even though Odent has been made a hero by many birth activists in this country, we would do well to realize that his views share a lot with those of nineteenth-century writers who relegated women to the "natural" realm of the domestic.

Instead of seeing the Pithiviers women as engaged in a "natural" lower-order activity, why can we not see them as engaged in higher-order activity? The kinds of integration of body and mind fostered by the psychophysiological approach and others, the kinds of wholly involved activity captured by the metaphors of the journey and the dance, could well be taken as higher, more essentially human, more essentially cultural forms of consciousness and activity. Here, perhaps, are whole human beings, all their parts interrelated, engaged in what may be the only form of truly unalienated labor now available to us.

If the fragmentation in the ordinary language we use to talk about birthing is in part a result of scientific medicine's definition and treatment of birth, what happens to that language when women arrange to birth in a different setting, according to different principles? The implicit imagery of separation—"the contractions come on," "I go through the labor"—is still commonly used, in part I am sure because its implications are so difficult to see. But women reach out for other ways of describing the experience, ways that assert wholeness in the face of being broken apart. One woman (Nancy Jankowski), who described her second home birth after a cesarean, told me of a difficult labor, with days of first-stage stopping and starting, a baby presenting posteriorly, hours of painful pushing, and finally the birth of a baby weighing over ten pounds. Yet there was no trace discernible in her words that anyone had considered anything wrong, that any special measures had to be taken, or that there was any cause for alarm. She, her friends, and midwives evidently saw birth as a process that would happen in its own time, as something she could do herself, and, although it was hard, as cause for an atmosphere like a party. The mother summed up the experience by saying, "I felt so whole"; when the baby didn't breathe immediately, the midwife blew softly in his face to show him what to do.

10 Menopause, Power, and Heat

I have heard women compare their sensations to burning steam rising from the pit of the stomach. These flushes may be considered as cases of pathological blushing.
—Edward Tilt
The Change of Life in Health and Disease (1857)

During their birthing years women find many rich ways of resisting medical models and developing alternative ones that reintegrate human experience. What happens during their later years? Is there an alternative "cultural grammar" of menopause despite its public image of atrophy and failure, a grammar that would belie these public images?

I begin with "hot flashes," the one true "physical" indication specific to menopause.[1] It must be said at the start that there is no good way of separating the purely "physical" aspects of hot flashes from their social and cultural context. Researchers have found marked variations in the incidence of this experience and in how it is regarded by women. In studies using comparable data-gathering techniques, 69.2 percent of a sample of Canadian women report having experienced a hot flash at some time, while only 20 percent of a sample of Japanese women report it.[2] Women in a Newfoundland fishing village see bleeding during "the change" as healthy, for it is the final purge or cleaning out of the body. "Flashes and flushes are caused by 'too much or bad blood' and [are] welcomed as purifiers."[3] Edward Tilt's mid-nineteenth-century account of "flushes" (*not* "flashes," but "flushes") as "compensating discharges" beneficial to the body hints that an earlier medical and perhaps popular culture may have welcomed "flushes" as evidence of strength, inner harmony, and balance.

The cultural grammar of hot flashes that we discovered in our interviews reveals something quite different from strength and inner harmony, accustomed as we are to considering menopause as a combination of broken-down central control and the end of production. When I began to read about menopause and hear what women said in the interviews, I had a feeling that my analysis of PMS was going to repeat itself. For example, consider these two accounts of hot flashes in popular health books:

> I was reading a book and suddenly this wave of heat came over me. I couldn't believe it was happening to me. I rushed to the mirror, and my face and neck were as red as a beet. It was *my* face, but I was so shocked and while I was looking and touching my face, it started to disappear and in a minute it was all gone. It wasn't really scary—but it was.[4]

> I once had one that lasted ten minutes as I was eating lunch one day. All of a sudden my face got suffused with heat and became all screwed up and twisted while my jaws got stuck so tight I couldn't eat another bite. When I looked in the mirror, I hardly recognized myself, I was so twisted and lined. But just as unexpectedly as it had come, the whole thing vanished.[5]

As in the case of premenstrual syndrome, women experiencing dramatic transformations in physical and emotional states are not able to recognize themselves. Could it be, I wondered, that women in menopause are experiencing the rage and anger associated with PMS? The more I thought about it, the more it seemed that the culprit involved in hot flashes, which appeared to women as something external taking over their selves, was not anger, but embarrassment. Consider Goffman's description of embarrassment: "blushing, fumbling, stuttering, an unusually low- or high-pitched voice, quavering speech or breaking of the voice, sweating, blanching, blinking, tremor of the hand, hesitating or vacillating movement, absent-mindedness, and malapropisms."[6] Not all these factors are present in hot flashes as women describe them, but many key ones are: turning red, sweating, and confusion.

In Goffman's classic study of what social situations give rise to embarrassment, he focused on events in which the self a person intends to project to others is confronted with another self he or she would rather not acknowledge or one that is incompatible with the present situation. "Because of possessing multiple selves the individual may find he is required both to be present and to not be present on certain occasions. Embarrassment ensues: the individual finds himself being torn apart, however gently. Corresponding to the oscillation of his conduct is the oscillation of his self."[7] Goffman has in mind situations such as when a top-level executive and a lowly secretary find themselves eating across the table in the same cafeteria.

In a recent review of further efforts to explain what kinds of situations cause embarrassment, Edelmann identifies several types: (1) Goffman's description of failure to confirm the self-image; (2) loss of poise or failure of social skill such as failure to control bodily functions, stumbling, or spilling things; (3) failure of meshing, such as when a person realizes that another person is more important than she thought or has a deformity she did not notice; (4) breaches of privacy, such as having one's personal space invaded or becoming the center of attention because of either praise or criticism; and (5) empathic embarrassment, as when one is embarrassed on realizing that someone else is or should be.[8]

Nearly all these kinds of interactions can be involved in hot flashes. Many women's comments relate to failure of meshing, in that their judgment about a room being too hot does not mesh with other people's. "You just ask, 'Is it warm in here?' Then all of a sudden you get real hot, and 'Uh oh there it is,' just like a wind, it comes and lasts maybe a minute" (Regina Hooper). "I can sit here and be real chilly and the next thing you know I'm coming out of my clothes. It's weird. And then you get chilly again. You go back to being chilly. Everybody thinks you're crazy!" (Martha Gibson).

Many women mentioned that the hot flashes were embarrassing, without being able to articulate why. "I can remember being very embarrassed talking with a fellow worker and my face turned very red. [This was during work?] I was sure that it was written all over my face what the cause of that redness was and there was nothing for me to be blushing about, nothing in the conversation that was embarrassing, so I remember those little embarrassing episodes. [How were you feeling?] Nothing, just praying that he didn't notice it" (Martina Ostrov). Insight into these remarks can be gained by using the analyses of embarrassment above: hot flashes make you feel like everyone is looking at you (number 4); they are an outward public sign of an inner bodily process associated with the uterus and ovaries, which are supposed to be kept private and concealed (numbers 2 and 4); they reveal indisputably that one is a woman and a woman of a certain age; in situations where one is projecting the aspect of the self as colleague, fellow worker, leader, or reliable functionary, these other parts of the self may be felt to be inappropriate (number 1), so much so that mentioning them brings on empathic embarrassment (number 5).

One woman's account incorporates almost all these elements:

> I am active, I am busy and thinking. That is when it would happen. But if you are talking to someone eyeball to eyeball and you are trying to convince them and sell them on something and you start sweating, they look at you. Of course if I was talking to a doctor they would understand. But you talk to someone and they say, "Why are you sweating?" [What did you say?] I think that in the beginning I used to say that I was hot and I can't take the heat or I just had a cold. And after a while I would say that I was going through my menopause, damn it. That is what you should say. That shuts them up fast. Especially a man. [Were they embarrassed?] Yes, well, let them be embarrassed. I was tired of being embarrassed. [Were you embarrassed?] Yes, I was embarrassed. (Gladys Sundquist)

Hot flashes and women's and society's responses to them are layered with levels upon levels of intentionality and interpretation, just as the same muscular contraction of the eyelid can be a twitch, a wink, a parody of a wink, a rehearsal of a parody of a wink, or a fake wink.[9] We have internal and external physical sensations (heat, breathlessness, flushing, sweating) that are similar to what happens when we are embarrassed; on top of that comes the immediate realization that others are not hot and that they may see all the outward signs of embarrassment and know the inner state—menopause—that means one should be embarrassed, even if one isn't.

One aspect of menopause is not elucidated much by the existing research on embarrassment: the effect of relative power and status on interaction. In empathic embarrassment, for example, the feelings of the person experiencing embarrassment because someone *else* has lost face or made a mistake may be similar in some ways to the other person's embarrassment, but he or she has in fact *not* lost face or made a mistake. Weinberg gives the example of a medical intern who is embarrassed when examining a woman patient's genitals.[10] He may well be embarrassed, but it is not the intern whose privacy is invaded, whose sexual parts are scrutinized and handled by a stranger. I suppose nearly any emotion can be felt vicariously—I can feel fear or terror when I see someone else in danger—but it is one thing to be the person threatened and another to be the person witnessing it.

This point is significant because of the greater number of occasions in which women are in subordinate positions that might increase their propensity to feel embarrassed. Further research could clarify this. At any rate, it is clear from my interviews and the literature that women associate hot flashes with situations where they are "nervous" or especially want to make a good impression: "Worse, an emotion-charged situation, such as addressing the chairman of the board, may bring it on. Women often complain that flushes always seem to creep up at the worst of all possible times. One patient exclaimed, 'It's hard to appear calm, cool, collected, and sophisticated when suddenly you turn red and break out in large drops of perspiration all over your face. It always throws me when it happens.'"[11] "Alida, 46, a dress designer, was in a conference showing her sketches. 'Suddenly I became a mass sensation of hot pricklies. No one said a word, I'm not even sure they noticed, because I was so excited about the designs.'"[12] "I noticed that I was having them mostly at work and not so much on the weekends. I attributed it to the fact that I have a lot more stress at work" (Ruth Carlson). These might be precisely the occasions on which a person in a more tenuous, subordinate position would have the most to lose, be the most apprehensive about her performance or appearance, and hence far more likely to feel the "hot flash" of embarrassment.

If women, generally in more subordinate positions than men, are likely to suffer embarrassment in connection with hot flashes, are some women more likely to suffer it than others? Although most studies of the incidence and perception of menopausal symptoms have involved only the middle class,[13] a few look at variation among classes. An early study found that working-class women are more anxious during menopause than after it, while middle-class women do not change in anxiety significantly as they move beyond menopause.[14] A more recent study based on questionnaires to ascertain women's attitudes toward menopause found that in general women do not have an illness orientation toward menopause. But there was variation by occupation in the extent to which women saw menopause as an illness: professional women had the highest wellness orientation, secretarial-clerical second, homemakers third, and blue-collar workers lowest.[15]

However incomplete, these studies still indicate that the direction of distress with menopause runs in the same direction as illness and death rates generally: the farther down the class hierarchy, the more suffering and perception of illness there are.[16] The most sophisticated study to date was done in Belgium and corroborates this general finding. Researchers on menopause usually trace hot flashes, which they regard as one of the most invariant accompaniments of menopause,[17] directly to physiological factors such as fluctuating estrogen levels.

They usually hold these factors to be independent of social conditions.[18] Yet the Belgium study shows higher socioeconomic classes having fewer hot flashes. Similarly, on broad indicators of nervous and vasomotor symptoms, the lowest incidence is found among the highest class. When this study measured general satisfaction with life and degree of alienation, it found they also correlated with class: higher classes are more satisfied with life and experience less alienation than lower classes. These patterns replicate the clear positive association between higher levels of general satisfaction with life and measures of greater health and longevity found in studies in the United States.[19]

None of this should be surprising. The 1973 Health, Education, and Welfare study on work in America found that the single strongest predictors of longevity were not genetic heritage, physical functioning, or use of tobacco but simply one's general satisfaction with work and overall happiness.[20] These might be important predictors for contentment during menopause also, where the process, even though it certainly has a physiological aspect, is often accompanied by signs of aging and changes in family composition, both of which are very differently interpreted depending on how much security and satisfaction one experiences in life generally.

To understand better the links between power, subordination, gender, and the underlying cultural grammar of hot flashes, consider that our imagery of power and control usually use both space and temperature metaphorically. As Lakoff and Johnson have shown,

HAVING CONTROL OF FORCE IS UP; BEING SUBJECT TO CONTROL OR FORCE IS DOWN
I have control *over* her. I am *on top of* the situation. He's in a *superior* position. He's at the *height* of his power. He's in the *high* command. He's in the *upper* echelon. His power *rose*.[21]

HIGH STATUS IS UP; LOW STATUS IS DOWN
He has a *lofty* position. She'll *rise* to the *top*. He's at the *peak* of his career. He's climbing the ladder. He has little *upward* mobility. He's at the *bottom* of the social hierarchy. She *fell* in status.[22]

And, key for our purposes:

RATIONAL IS UP, EMOTIONAL IS DOWN
The discussion *fell to the emotional* level, but I *raised* it back *up to the rational* plane. We put our *feelings* aside and had a *high-level intellectual* discussion of the matter. He couldn't *rise above* his *emotions*.[23]

I would add to these metaphors a contrast between hot and cold:[24]

RATIONAL CALCULATION IS COLD; EMOTION, ESPECIALLY ANGER, IS HOT

She is calm, *cool,* and collected. It was a *cold,* calculating move. He quit *cold* turkey. It was a *cold*-blooded murder. She shed the *cold* light of reason on the matter.

I lost my *cool.* I blew my (*smoke*) stack. She had to let off *steam.* It *burns* me up. Her letter was so angry it *burned* up the paper. She saw *red.* He was *inflamed;* she was *incensed.* "This kind of anger would *flare up* and *burn* the innocent."[25] She got *hot* under the collar. I was *boiling* mad.

Taken together, there is a certain systematicity, as Lakoff and Johnson term it, among these concepts, such that power, height, rationality, and coolness go together on the one hand and lack of power, low position, emotions, and heat go together on the other. It is fairly obvious where this leaves women experiencing hot flashes during menopause, whether the precipitating circumstances, if any, promote embarrassment or anger: it leaves them hot and bothered, down and out.

Perhaps the element of subordination in relationships may lead to hot flashes and embarrassment in situations where subordination might otherwise lead to anger. Consider the case of a woman whose severe night sweats and hot flashes were completely stopped for three weeks after she decided to try taking vitamin E on her own. "A month later, after three weeks without a flash, Priscilla went to see her doctor and told him about this miraculous change. He said it was nonsense. On the spot, that instant, Priscilla had a huge hot flash."[26] No doubt she did have a huge hot flash, but if she had not been in menopause, might she not have simply said she was furious at being so patronized?

Or consider another woman who began to sort out the overlapping layers in her experience of hot flashes:

> It is frustrating because they look at you in amazement. I would just perspire from the top of my head and it would just drip down over my eyebrows. All over my face. The rest of my body was dry. It was my head. In hot weather it is from my head. It drips down into my glasses and so on. It is miserable. I can remember many occasions. [As someone who hasn't had any yet, the interviewer wonders what it feels like.] It is anger. I am angry. I am angry that my body is doing this to me. I think that that is it. I am angry. I am annoyed. I didn't mind if I was by myself but it didn't happen as much when I was by myself. It was when I was more tense or not even tense, it would happen at the craziest times. (Gladys Sundquist)

In Chapters 3 and 4 I discussed the dominant metaphors in medical discourse about a woman's body—as a hierarchical, bureaucratically organized system under control of the cerebral cortex and a manufacturing plant designed for production of babies. We considered how menopause is seen in the terms of this model as a case of the breakdown of authority: ovaries fail to respond, and the consequence is decline, regression, and decay. At this point, I turn to how women themselves experience and describe the process.

Recent research on menopause has shown repeatedly that despite the way this event has been heavily described in medical terms, for most women it does not present a major problem.[27] This was true for most women we interviewed. But the way they express their responses to menopause is extremely significant. Over and over, women describe menopause by saying, "It was nothing." "Nothing. Never had any problem, it just stopped, it slowed up" (Ruth Chapman). "Nothing. Just stopped and that's about it" (Freda Von Hausen).

And from younger women: "It just stops. There's childbearing years and when she gets older her body isn't able to bear children. It just kind of stops" (Lisa Miner). "Mom has just gone through it; it was like a light going out. It just sort of stopped" (Mara Lenhart). "You have your period for ages and all of a sudden it stops" (Julie Morgan). "My grandmother stopped menstruating when she was thirty-five. From what I can tell, that's all. Stopped and absolutely nothing" (Leah Rubenstein).

Menopause, for women who have no particular trouble during it, is described as the cessation or absence of something—menstruation—which leaves them with nothing. Although this sounds like the production model, in which menopause is seen as female organs failing to produce, there is no indication that women, middle-class or working-class, see this in negative terms. No one used the terms common in medical parlance: ovaries failing to respond, hormones' production declining, decreasing sensitivity of the hypothalamus, and breakdown of the feedback system. Even when asked how they would explain menopause to someone else, women repeated, "It stops." They are simply saying that menstruation was there and now is not. Later we will see whether they miss it.

What about the metaphor of a hierarchical organization? Do women see this period as one in which their body parts get out of control? It is striking that none of the women past menopause described it in terms like these. Only two women even used the term "control," one to say that an estrogen compound prescribed by her doctor had brought the symptoms she experienced under control and another that surgery had controlled her excessive bleeding.[28] In startling contrast, a great many of the young women we interviewed, looking ahead, saw menopause as a time when one's body is out of control. "I guess it's more of a fear, not of post-menopause, just the actual process. You're not really in control of your body. That much is not predictable, that's what scares me about it" (Tania Parrish). Describing her mother's reaction to hormones prescribed by a gynecologist, another woman said, "Her body temperature is a lot more controlled" (Rachel Lehman). "The unknown of menopause, that out-of-controlness. That's one of the biggest things we have to change if the world is going to be sensible and civilized for women to live in. As with everything else, we feel out of control with that. I can't imagine not feeling out of control, because biologically it seems that these things are out of our control, but some of it is a social overlay too" (Meg O'Hara). "My grandmother almost went insane, she almost didn't make it through menopause at all" (Marcia Robbins). "Mom almost went berserk. I don't mean she really did, but it was hard" (Ann Morrison). "Watching my mother go through that. I think that is why she is kind of whacko" (Gina Billingsly).

There are a number of possible reasons for this discrepancy between young women and menopausal women. It could be that young women perceive their mothers' behavior as out of control just because children generally fear any unusual behavior in their parents. Or it could be that the younger women have more thoroughly internalized the medical model of hierarchy and control.[29] Watching their mothers and grandmothers go through menopause, they share with some medical practitioners the interpretation of older women's behavior as being out of control: "vacillating and often irrational."[30] Some of these descriptions in medical terms deserve quoting at length.

> Mrs. M., normally a fairly easy-going person, found herself waking up most mornings with frustrated, rebellious feelings. Somehow the usual habits of her family had become maddening.
>
> [The author describes a mutually irritating exchange between Mrs. M. and her husband.]
>
> Their daughter rushed through the kitchen saying, "Hi, Mom, 'by Mom" and grabbed her jacket from the hall closet. She knocked down a coat and two sweaters in the process.
>
> "Now see what you've done—come back here and pick them up," commanded Mrs. M.
>
> "Look Mom, I'm late, I'll miss the bus!"
>
> Suddenly Mrs. M. felt herself losing all restraint. Words poured out of her, bitter hateful words.

"Giving you the best years of my life! And you can't do a simple little thing for me. Of all the ungrateful . . ." The tirade went on and on.[31]

The author citing this example goes on to describe the woman's behavior as being "unable to meet even minor problems. Mrs. M. couldn't even cope with her daughter's untidiness." I would reply that her behavior seems entirely appropriate, especially assuming that she had done years of Sisyphus-like housework and was now finding herself rebelling against being a servant for her nearly grown children. But it is not just that Mrs. M. is in trouble. The author worries most of all about the effect menopausal women have on their children:

> Unfortunately, the adolescent's personality often parallels the mother's in being egocentric, vacillating and often irrational. However, for the adolescent, this behavior is expected but not for a mature, slightly past middle age woman.[32]

Another medical practitioner agrees that the characteristics of menopause are unfortunate but sees their effects rebounding onto women themselves:

> I cannot help feeling that the reason so few women being [sic] found in leading positions is to be at least partly explained by the mental unbalance in these years around the time of the menopause. It is around the age of 50 that men take the final step to the top, a step that women with equal intellectual capacities rarely take. I know that many aspects are involved, but the climacteric may well be an important one.[33]

In sharing the perspective that women in menopause are out of control, younger women may be unwittingly perpetuating a bias that when women step out of an accustomed—even if no longer wanted—role and protest, resist, or act in the world, they are defined as sick and weak (just as women with PMS are).[34]

Certainly the vast majority of the older women we interviewed saw menopause in a positive light. It meant pleasure at avoiding whatever discomfort they felt during periods and relief from the nuisance of dealing with bleeding, pads, or tampons: "I was pleased that I didn't have the inconvenience of the menstruation any more. I was very happy to arrive at the menopause" (Barbara Heath). For those women sexually active with men, it meant delight to not suffer the fear of pregnancy. This last concern was expressed in vivid terms since for most women in this age group, abortion was illegal during their reproductive years: "I was glad. It was definite that I wouldn't have to worry about pregnancy. I had my last child at forty-six, and I certainly didn't want to rack up any record for having a baby at fifty-five or sixty" (Claudia Williams). Also on the positive side, many women asserted that menopause had meant no feeling that they had lost their womanhood. "I sure didn't feel deprived or that I was losing my youth; it really didn't bother me one bit. I always had bad cramps, I felt relief from the discomfort, I didn't sit and weep for any lost womanhood" (Estelle Hoffman).

My research assistants who did interviews with older women often commented that it was difficult to keep them on the subject of the interview because they wanted to wander from menopause to talk about many other aspects of their lives. Taken as a whole, it seems clear that these women do not experience menopause as if it were a separate episode in life akin to a stay in the hospital for an illness. They describe it as a part of all the other events happening in their lives: "It was just part of life. I had reached that stage. That was another phase that had passed. I was glad of it. I wasn't going to have to worry about it anymore" (Eleanor Pittman). "So all that [trouble in relationships with men, her son getting arrested many times freedom-riding in Mississippi] was mixed up with the menopause" (Ethel Jacobsen).

For some, menopause was a milestone that led them to take stock of their lives and reach for greater happiness. "I realized that, boy, I've reached another milestone. If I was going to do anything, I'd better do it. And I'm talking now about our marriage. When I told my husband I wanted a divorce, he couldn't believe it. He thought I had lost my mind. You can't help but realize that time is going by, you know. And I was healthy, so I figured I wasn't going to drop dead" (Claudia Williams). In this case, her husband saw her as having lost her mind; she saw herself as having changed: "The way I see it I changed and he didn't. And I couldn't understand why he didn't." This suggests that younger women's perception of their mothers and grandmothers as being out of control when they went through menopause might not be matched by those mothers and grandmothers who see themselves as taking new steps toward independence, strength, and power.

For at least some middle-class and working-class women, reaching the other side of menopause can mean greater feelings of energy and strength.

> I tried to work up some nostalgia because you're supposed to, but I really didn't. Because I did not want any more children, what kind of insanity is this that I would make a big deal about it, I didn't want any more children. I got a job. It's wonderful to have your own life. Money is power. To be able

to control your own life, to be independent, take care of yourself as you need to gives you a great power. (Estelle Hoffman, supervisor in a government agency)

[How did you feel after you finished menopause?] It was just like somebody injected me with strength and energy and enthusiasm. I started doing more things. Of course my job had bad hours, so I couldn't do a lot of things. (Ethel Jacobsen, retired garment trade worker)

The general cultural ideology of separation of home and work appears in this material when women are embarrassed at having their menopausal state revealed publicly through hot flashes. As with the hassle of menstruation, women are asked to do what is nearly impossible: keep secret a part of their selves that they cannot help but carry into the public realm and that they often wear blatantly on their faces. Resistance in the case of menopause does not consist, as it does in the case of menstruation, in turning private spaces to seditious purposes. It consists in the occasions when women publicly name their state, claiming its right to exist as part of themselves in the public realm, and embarrassing their male coworkers at the same time, paradoxically producing in the men our cultural emblems of subordination: heat and emotion.

Resistance to the medical model runs along a generational line. Although many younger women share the medical view that menopausal women are out of control, women going through menopause by and large do not see it this way, but instead see it as a release of new energy and potentiality. Some women, as described in this poem, even manage to harness the anger provoked by their position in society to their desire for a different kind of life.

Mid-Point

She stored up the anger
for twenty-five years,
then she laid it on the table
like a casserole for dinner.

"I have stolen back
my life," she said.
"I have taken possession
of the rain and the sun
and the grasses," she said.

"You are talking
like a madwoman,"
he said.

"My hands are rocks,
my teeth are bullets,"
she said.

"You are
my wife,"
he said.

"My throat is an eagle,
my breasts
are two white hurricanes," she said.

"Stop!" he said.
"Stop or I shall call
a doctor."

"My hair
is a hornet's nest,
my lips
are thin snakes
waiting for their victim."

He cooked his own dinners,
after that.

The doctors diagnosed it
common change-of-life.

She, too, diagnosed
it change of life.
And on leaving the hospital
she said to her woman-friend
"My cheeks
are the wings
of a young
virgin dove.
Kiss them."[35]

Consciousness and Ideology

11 Class and Resistance

Thus the ideological element in human thought, viewed at this level, is always bound up with the existing life-situation of the thinker. According to this view human thought arises, and operates, not in a social vacuum but in a definite social milieu.

We need not regard it as a source of error that all thought is so rooted. Just as the individual who participates in a complex of vital social relations with other men thereby enjoys a chance of obtaining a more precise and penetrating insight into his fellows, so a given point of view and a given set of concepts, because they are bound up with and grow out of a certain social reality, offer, through intimate contact with this reality, a greater chance of revealing their meaning.

—Karl Mannheim
Ideology and Utopia

So far we have come across several kinds of consciousness of oppression or resistance to it: the often inchoate anger associated with PMS and menopause, the silence of working-class women on scientific views of menstruation, the many ways women try to resist the imposition of the production model during birth and strive to create different meanings. In this chapter I want to bring together the question of consciousness and class to ask a few limited questions about the relationship between the two.

The literature on ideology and class contains two major opposing positions. Put simply, the first argues that as one goes down the hierarchy of class, race, and gender, the possibility of an alternative consciousness or resistance to the status quo diminishes; the second argues that as one goes down the hierarchy, the increased oppression can lead to greater consciousness and resistance to the status quo. For example, Juliet Mitchell attributes the dominance of the feminist movement by the middle class to the fact that "the most economically and socially underprivileged woman is bound much tighter to her condition by a consensus which passes it off as 'natural.'"[1] On the other side, Leith Mullings argues that greater oppression leads to greater consciousness and resistance: "Yet, being triply oppressed, minority women are also a triple threat. With their consciousness shaped by their experiences as workers, as members of a minority group, and as women, they are at the core of resistance."[2]

The issues here are complex. Mitchell is claiming that the woman at the bottom is silent because she does not see her oppression, Mullings that the woman at the bottom not only sees her oppression but both speaks out against it and organizes against it. Between these two extremes are many possibilities: women may know their oppression but choose not to speak out about it, judging the risk to be too great. They may do nothing or, operating stealthily in the interstices of power, they may resist through devious ways of speaking or acting.[3]

The logic of the position taken by Mitchell is similar to that found in the literature on class consciousness in what Abercrombie, Hill, and Turner have called the strong form of the dominant ideology hypothesis:

> In all societies based on class divisions there is a dominant class which enjoys control of both the means of material production and the means of mental production. Through its control of ideological production, the dominant class is able to supervise the construction of a set of coherent beliefs. These dominant beliefs of the dominant class are more powerful, dense and coherent than those of subordinate classes. The dominant ideology penetrates and infects the consciousness of the working class, because the working class comes to see and to experience reality through the conceptual categories of the dominant class.[4]

In other words, if the very categories of thought are controlled and determined by the dominant class and formed in its interest, to the extent this thought penetrates the whole society no one can think in other categories. This same argument has been applied to the situation of women: "The world simply was and we were in it. We could only touch and act upon its outer shapes while seeing through the lens men made for us. We had no means of relating our inner selves to an outer movement of things. All theory, all connecting language and ideas which could make us see ourselves in relation to a continuum or as part of a whole were external to us. We had no part in their making . . . But where was an alternative consciousness of ourselves to come from?"[5] "The tool for representing, for objectifying one's experience in order to deal with it, culture, is so saturated with male bias that women almost never have a chance to see themselves culturally through their own eyes."[6]

In addition to clearly stating the dominant ideology thesis and showing how thoroughly it has permeated contemporary writing on consciousness, Abercrombie, Hill, and Turner present an alternative hypothesis that allows much greater room for the working class to resist. They argue that the criteria for what counts as resistance have been held at an unreasonably stringent level and that researchers have not been looking for resistance to dominant views in the right places.[7] It seems a terrible irony that although Abercrombie, Hill, and Turner have developed a view that gives greater credit and dignity to the working class, they do not extend this to women. They do not take up gender as an issue, but when they mention women explicitly under a discussion of feudalism, they assume that women were *controlled* by Christian ideology.[8] What is sauce for the gander ought to be sauce for the goose: if ideologies are not all-powerful in constraining those they potentially confine, why should they constrain women more than men?

Others are also searching for ways to avoid assuming the existence of blinders on those at the bottom of social systems. Göran Therborn puts "subjection-qualification" at the center of how ideologies function: this refers to the dual and simultaneous process by which humans become qualified to serve social roles and gain the capacity to modify the social roles themselves. At the heart of people's ability to question the social order is their ability to conceive of an alternative kind of regime. If they cannot conceive of alternatives, they will have a sense of inevitability about what is, a sense of deference toward present rulers, and a sense of resignation about what is possible. If they can, they may begin to envision some form of opposition. Therborn describes only the beginning wedge of opposition, however: held in place by fear, people may oppose and disobey but are not apt to systematically push until their demands are satisfied.[9]

This is a useful start. Returning specifically to women, and considering all our interviews with these questions in mind, it strikes me that there are a great many ways women express consciousness of their position and opposition to oppression. Below I list some of the many forms consciousness and resistance take. Any such list could never be exhaustive. These are only meant as a few of the brightest and most easily visible lights in the firmament.

Acceptance: Things just are as they are; nothing can or should be changed.

[How would you react if you never had to menstruate again?] That would be nice. But I know that's life, that's what every woman has to go through. There's no way out. [If someone magically offered that you wouldn't have to do it for the next thirty years, would you take the offer?] No, I'd just go ahead and let it be, because sometimes things have to be this way, even though you don't want them that way. You have to live with them. [Can you suggest changes in society that would make it easier?] No, not really. (Janice Peterson)

[Did you ever have any physical changes during menstruation?] Yes. Just cramps. I felt that when I grew up that went with the territory. I knew that all my sisters had cramps and thought it was awful. I don't know if mine weren't that bad or if I just accepted them. Because nowadays everybody has PMS. When I was growing up, you just had it. There wasn't that much to it, you just suffered in silence because you were supposed to. (Teresa Cresswell)

[Is there anything in very general terms that you would say might improve women's condition?] I don't know what to say about improving life. Each one of us has a different life to lead and different expectations of that life. And what might improve my life might not improve your life, it might not make you happy. [What would improve yours?] I don't know. I haven't given it any thought. I just take things as they come along and enjoy them. I am sure that every day something new will be discovered and there will be new ways of using it. (Edna Summerdale)

Lament: A focus on grief, pain, or unhappiness, with or without perception of *structural* factors outside the individual's control. Laments may or may not go with a conviction that things could be changed; their tone may be anything from self-pity to righteous anger.

[This young woman has been considering whether she would give up menstruating if she were offered the chance. She decides she would not because she definitely wants to have children. The interviewer then asks: Is there anything in general you would like to change about our society?] It's weird how they commercialize the whole thing, into competition in business for pads and stuff. Different ones that do everything possible. It's ridiculous, the commercial aspect is awful. People are making money off a natural process. It's not like buying clothing, it's an added expense that women have. I don't see that there's any way to change anything. No matter what, people are still going to buy the products. (Linda Ansell)

[If you could change anything?] I think it would be a little bit easier if a lot of women got together and sat down and talked. More than just one person at a time. Try to figure out what everybody is going through. I know that next time I am going to be more choosy about who I am going to go with, and who I am going to have a kid by. Because there is no way I am going to put my son through that again. He has lost his father and Billy [the man she last lived with]. (Patricia Barton)

[Would you like to see any changes in our society?] Some women have cramps so severe that their whole attitude changes, maybe they need time to themselves and maybe if people would understand that they need time off, not the whole time, maybe a couple of days. When I first come on I sleep in bed a lot. I don't feel like doing nothing. Maybe if people could understand more. Women's bodies change, you know. (Linda Matthews)

[How is your health these days?] I have osteoporosis. They told me that about three years ago. I am drinking more buttermilk and milk and am taking calcium supplements. They knew all of these years about it [osteoporosis], why the hell haven't they done any research on it? Now they are beginning to advertise what women can do. The American Dairy Association is saying that women should drink more milk. How come that came only last year for a disease that only affects old women [osteoporosis]? They have never done anything about it. Now if there is discrimination against women it is that. So I am doing exercises. I went to a doctor and he said just be careful, there is nothing you can do. Doctors, I don't think that they even spend two minutes on osteoporosis. There are no specialists on it. There is nothing about it and these are older women. And it makes me furious and that is what I mean by discrimination. Discrimination against women. That is what it is. If men had that, let me tell you. They have done absolutely no research about it. None. (Gladys Sundquist)

Nonaction: Not participating in an organization, not attending a clinic or not using a term because it is perceived to be against one's interests.

[This woman has two young children and says she would like to have two more about seven years from now. The interviewer asks: Can you imagine those births?] I definitely won't go into the hospital until I absolutely have to. [Why?] Because when you're in the hospital, the time goes by a lot slower. I think it does. And you can only have one person in there usually unless you go to a birthing room. But when you are at home you don't have anyone telling you to stay in one place and to do this and to do that; you can boss everyone else around. And that's how I like it. I would definitely stay home until the very last minute when they're five minutes apart. (Marcia Robbins)

[She is describing the different experiences she had giving birth to her four children.] I had learned my lesson to stay home and move around, instead of at the hospital where they tuck you in bed and you can't move. The third one, I remember, I was so uncomfortable and the nurse would leave me in the room and I would lean over and crank the bed up, because they wanted you to lay down flat. Well, they caught me cranking the bed up and they wouldn't let me out of bed. So I learned to stay home with the fourth one. I stayed home until about one o'clock in the afternoon. And then I called my husband at work and said I can't wait any longer, I have to go. I found someone to stay with my other three children and I went to the hospital. He got there at about 4:20 and I had the baby at about 4:30, so it wasn't too long that time. I had learned to wait that time instead of sitting in the hospital. (Teresa Cresswell)

[This woman gave birth to seven children. By the time she had had five of them, she tells the interviewer, she had learned some things.] I got to the point where I wouldn't go to the hospital early, where I waited as long as I could because I didn't want to lie around. [What was it like to be lying around in there—lonely?] Yeah, you're alone, you're in pain, they come back and check you and stuff. But I'd rather be walking around at home. I stayed home and did things that day, so when I got to the hospital I didn't have to wait a long time. So, see how you learn things as you go along! (Gracie Evans)

[Having taken estrogen prescribed by one doctor for a month, this woman decides to consult another doctor to make sure she is all right.] He said, "It looks all right but I want to be careful and give you a D&C." So I left the office and called back and said that I had postponed it. I didn't have the time. Everyone that I know who has had a D&C where they scrape the womb, these women sooner or later get a hysterectomy. [Why do you think that was?] Well, that was one way of getting women under control. In the fifties it was a very reactionary period. It wasn't only reactionary politically but I can remember reading all of the articles saying that women shouldn't go to college and be smarter than their husbands. If you are smarter than a man you will never get a husband. There was a lot of propaganda. The hysterectomy was a way of punishing women. In an indirect way. I'm not saying that the doctor said I am going to punish you. It was a way for him to make money. (Gladys Sundquist)

Sabotage: Action or words meant to foil some process or behavior perceived to be detrimental but intended not to be detected.

Many women report surreptitiously eating and drinking during labor, removing monitors when nurses are not around, or going for long walks so the monitor cannot be used. One woman, who told her story in Chapter 8, was so determined to stall going to the hospital to avoid another cesarean section that she repeatedly concealed from her husband, family, and doctor what was going on.

I wasn't letting them know that during a contraction I couldn't do much. I could talk a little bit but I didn't want it to catch them. I could tell when it was coming on. I'd go, "Excuse me" and go to the bedroom. Got to where Friday I was in there hanging on the dresser going [breathes while groaning]. (Janice Sanderson)

Resistance: Refusing to accept a definition of oneself and saying so, refusing to act as requested or required. This can be done as an individual or in concert with a group.

In high school we had sex education and I remember writing a letter to the editor of the *Washington Star,* which got published, saying that you better teach us this stuff, because if you don't that is not going to stop us from having sex, so you might as well tell us what we are doing. It included a lengthy quotation from *Our Bodies, Ourselves.* (Carla Hessler)

[This woman's baby was born six weeks prematurely. After she was able to bring him home, he stopped breathing and turned blue. The baby was taken back to the hospital to be checked for heart trouble.] I asked the doctor if he was a high risk SIDS [Sudden Infant Death Syndrome] baby. And he said, "What if I tell you no?" I said, "Then you would tell me no." He said, "What if I told you yes?" I said, "Then you would tell me yes. I am not going to flip out." So he said, "Whatever answer I tell you won't be helpful, so I am not going to give you one." And he walked out of the room. It took everything I had to not knock this man in his teeth. I was that mad. He was another resident. I said that I am getting out of this teaching hospital if it kills me. This child is not going back to a teaching hospital. So I wrote a letter to his pediatrician telling him that I didn't want him in that teaching situation. (Becky Sokolov)

[Her first child was born by cesarean. She is describing the vaginal birth of her second baby, in which she has had no anesthesia.] I had this argument with the doctor because he wanted to do an episiotomy and he wanted to give me a local. We had heard something about even a local anesthetic getting into the baby's bloodstream. And I didn't want him to do it. And he kept wanting to give me the local and I said just do the episiotomy without the local. He was really insistent. The chief of obstetrics was standing right at his elbow and I was sure that I was making the poor guy nervous. Finally the doctor said don't even do the episiotomy. Just leave her alone. So she was born and she was six pounds and nine ounces. She had tiny everything, and I tore maybe a quarter of an inch. (Sue Jackson)

[After this woman had three children she was refused a tubal ligation. She then had three more children.] Then they wanted to give me the tubal ligation. I told them no. You are not cutting me now. I will take care of myself from here on now. They tried. They gave me a hard time when I had the last child. What they really wanted was to give me a hysterectomy. I didn't want a hysterectomy, and every time I went to the clinic it became like a trauma because all of those doctors would come in and talk to me. They said that I had sugar. And they gave me all of these blood tests. They just thought that it was time for me to have one since I had six kids. They told me that I had sugar and that scared the daylights out of me. But I still wasn't going to have a hysterectomy. I was thirty-four or thirty-five and I didn't want them cutting up on me, because then you get sick, and I had never been sick. In fact, I read my chart and there wasn't anything on that chart that said anything about sugar. They accidentally left it on my bed one day and I saw it. And all of the blood tests that they had done were on there and there was no sign of sugar. So they put me through all of that just because they wanted me to have a hysterectomy. (Gracie Evans)

Sometimes the act of resistance is formed in fantasy afterward:

I remember when I went to the hospital. Considering I'm not in a good financial position, not being married and having a baby, she said to me, "Oh, you're going to put the baby up for adoption?" I really, really was angry. I felt like saying, "How dare you talk to me that way!" I didn't say that, but I really was very hurt. If she had to say it, she could have said, "Are you prepared to have this baby?" I felt like saying, "How dare you talk to me that way! I'm not some sleazo or some piece of shit or something!" (Elizabeth Larson)

And sometimes the odds against resisting successfully seem too great:

He did an episiotomy without asking or telling, he just did it. By that time I was his as far as he was concerned. I think really, we were so worn out. I knew he was doing an episiotomy when he started but I felt like what the hell. Just get the baby out. (Carol Gleason)

They didn't let me stay in labor too long. My doctor said he didn't want me to go through pain. [What did you think of that?] I wanted to go through it all, it wasn't hurting that bad. [So what would you have done if it was all left up to you?] I would have waited a little while longer. Because I was young, I didn't say anything. I just let him go ahead and do what he wanted even though I didn't want a cesarean. (Juliet Cook)

Rebellion: Forcing or persuading other people to change the way they talk or act, beyond the single instances of resistance like those described above. For example, women might organize and demand that a hospital (or obstetrics as a profession) change its policies for routine administration of anesthetic or performance of repeat cesarean sections. There have been many instances of successful rebellion in both the women's movement and the women's health movement. As a result of many individual acts and many activities of organizations, doctors and hospitals have changed their practices.[10] Because our interviews were for the most part not with health activists, and we did not ask those who were activists specifically about these activities, the most extreme form of opposition described in the interviews is generally "resistance" rather than "rebellion."

I give these illustrations only as a small indication of the range and variety of opposition apparent in our interviews. I turn now to the question posed at the beginning of this chapter, How does resistance of whatever kind relate to class? First I must say that my data may be somewhat weighted in favor of the portion of the population relatively more disposed to speak out or act in protest. For one thing, we always asked for volunteers from small groups, and even though we did not say the study had to do with these matters, women self-selected to talk about themselves may have been the most conscious and resistant ones. For another, women in the groups we contacted had often taken the step of joining a group or class, or at least seeking a service outside the home. This in itself may be an indication of relative activism. As a result of this probable bias, my data cannot be taken to indicate the level of resistance or opposition in the general population. However, there is no reason to suspect that our interviews with middle-class women were more likely to be with activists than our interviews with working-class women. Consequently my data can speak to the question of variation in consciousness or resistance by class.

To give an extremely rough measure of the degree of consciousness and resistance as it varies by class, I will look at a subset of the interviews, composed of women between 18 and 25 years of age. Restricting the subset to these younger women increases the comparability of their experiences: older women have more divergent lives, one having lived in California for a number of years, another having just moved to Baltimore from the deep South. Again for comparability, I selected twenty middle-class women in private college or high school and twenty working-class women in community high schools or colleges. Using the definitions of forms of consciousness and protest above, I read through the transcripts of the interviews, marking any section I judged to represent acceptance, lament, nonaction, sabotage or resistance.

Impressionistic as such a tally must be, I was struck by the extent of protest or resistance in the working-class group. Three middle-class women and three working-class women registered only acceptance. Six middle-class women and seven working-class women registered only laments with no sabotage or resistance. And nine middle-class women and nine working-class women reported some act of resistance. There were no incidents of sabotage and two incidents of nonaction (staying away from the hospital until the last minute), both working class. Added to this is the pattern discussed in Chapter 6 that only one of the working-class women made any reference to the scientific model of menstruation in describing how she would explain it to a young woman, while all of the middle-class women described the main medical metaphors of waste and decay. If this can be categorized as a kind of nonaction, rejection of a system detrimental to women, then the preponderance of resistance goes to the working class.

In discussing the relationship between class position and consciousness, Alison Jaggar summarizes the argument that the standpoint of the oppressed is more critical of existing society, more impartial, and also more comprehensive than that of the ruling class.

> Because their class position insulates them from the suffering of the oppressed, many members of the ruling class are likely to be convinced by their own ideology; either they fail to perceive the suffering of the oppressed or they believe that it is freely chosen, deserved or inevitable. They experience the current organization of society as basically satisfactory and so they accept the interpretation of reality that justifies that system of organization. They encounter little in their daily lives that conflicts with that interpretation. Oppressed groups, by contrast, suffer directly from the system that oppresses them. Sometimes the ruling ideology succeeds in duping them into partial denial of their pain or into accepting it temporarily but the pervasiveness, intensity and relentlessness of their suffering constantly push oppressed groups toward a realization that something is wrong with the prevailing social order. Their pain provides them with a motivation for finding out what is wrong, for criticising accepted interpretations of reality and for developing new and less distorted ways of understanding the world. These new systems of conceptualization will reflect the interests and values of the oppressed groups and so constitute a representation of reality from an alternative to the dominant standpoint. The standpoint of the oppressed is not just different from that of the ruling class; it is also epistemologically advantageous. It provides the basis for a view of reality that is more impartial than that of the ruling class and also more comprehensive. It is more impartial because it comes closer to representing the interests of society as a whole; whereas the standpoint of the ruling class reflects the interests only of one section of the population, the standpoint of the oppressed represents the in-

terests of the totality in that historical period. Moreover, whereas the condition of the oppressed groups is visible only dimly to the ruling class, the oppressed are able to see more clearly the ruled as well as the rulers and the relation between them. Thus, the standpoint of the oppressed includes and is able to explain the standpoint of the ruling class.[11]

Although this study was not designed to allow a thorough investigation of these claims, it can shed some light on them. The subset of the interviews analyzed above showed that working-class women are at least as likely to express opposition as are middle-class women. Another subset of the interviews gives an indication that we can in some cases go beyond that claim to say working-class women are more likely to envision the need for fundamental changes in society. I selected a subset of interviews that consisted of all the black women, middle-class and working-class. Along the lines of Jaggar's suggestions, this was to increase the degree of oppression experienced by the working-class group as much as possible. Within this subset, I considered only women's answers to the question, asked toward the end of the interview, *"Do you think there is anything that would make women's lives in general easier or better than they are?"*

Plenty of middle-class women and plenty of working-class women in this subset responded to this question with only flat acceptance of things just as they are. For example:

[21-year-old junior in college. Her father is a teacher, her mother a librarian.] I don't know. I don't find being a woman really that difficult except for the runny stockings and the high heels. I really couldn't tell you. Well, the fact, maybe that we have a period. I'll take that away. I guess that's what makes us different. If we gave him one, we might as well give him the boobs, too. [You like that difference?] Well, of course we have to be different from the males. It's just the way the world was created. God made man and woman, male and female, for us to reproduce. I just believe the human body is made the way it's supposed to be. (Nancy Parsons)

[19-year-old junior in college. Her mother does home care for the elderly, her father is assistant manager of a store.] What needs to be changed? I always say why couldn't men have half of our problems. But I wouldn't want to change that. That would go against God's word and what he put men and women on earth for. I don't think that I would change any of it with the exception of the expense, I guess they figure that women need tampons and Kotex and all of that so let's charge them a lot. (Linda Ross)

[27-year-old part-time nurse's assistant.] I can't change nothing. [If you could?] Don't have no pains, because that pain hits you real hard. No labor pains either. I guess that's it. I like being a woman, I just love it. (Bernadette Jones)

However, it is emphatically the case that the most extended, acute, definitive statements about the need for systematic changes all came from the working-class women in this subset. For example:

[28-year-old nurse's aide with one child. Her mother is a housewife, her father a construction worker.] Well, I think that I would change the way that the system is set up as far as how you are going to pay for all of this. That is a big hassle if you are just working part time. You don't know where your money is going to come from. You don't want to go begging someone for it. But you have to do something. And if you make a certain amount it is too much. I don't think that they take into consideration that you have to eat. I think that certain medical things should be taken care of. It shouldn't be a big hassle as long as you can show that you are not trying to get over [take unfair advantage of the system], so to speak. The United States is rich enough to take care of a lot of medical things. Female, male problems. It would be a lot easier all around. I don't see why we can't. (April Hobbs)

[54-year-old clerk in a small grocery store, divorced with seven children. Her mother and father did farm work in the South.] Well, I would like to see more education in terms of women taking care of themselves. Or knowing that they are going to have to or may have to take care of themselves at one point in their lives. And I would like to see women educated in that they don't have to be servants for men in order for them to meet with their approval. I guess that I am talking about equality. The word *equality* is not specific enough. I would like to see women more accepted as they are. I see women putting themselves out so much to be something, and I guess that that is the general way that society goes. But women are really great people. They take care of everything and they don't get any credit for it.

I think that we need to come together more. White women are nowhere near each other. They are so far apart. That is the way things are in society. They keep you divided, then you can't make much progress. Someone is always saying, "Well, I am better simply because of this." There are a lot of things that are generally happening to black and white women. It doesn't make any real difference. The economic situation for women, for black women, and I am sure for white women—they have economic situations too, not many because most of them are rich, right—but the economic situation for women is hard. I find that I have had an awfully hard time economically. (Gracie Evans)

[25-year-old security agent and domestic worker.] It is not just the men that are so biased against women, it is women themselves who are even harder on women. If I had my way I would try in some way to change that. Because even in the Bible it speaks of women as being childbearers and taking care of the home and that type of thing; I think that is one of the big reasons why society views women in that fashion. Even women themselves. They don't give themselves enough credit for being intelligent, intellectual human beings. A woman is not supposed to drive a bus and she is not supposed to be a construction worker. They should do things like I do, be a maid or something, but don't do anything physically exerting. [Can you think of anything that would help to bring about those changes in women?] It would have to be started on a small scale, of course. The only thing that I would do would be to get together with groups of women and have rap sessions. It will be almost impossible for women to change their minds about themselves because religion plays such a big part in society. Religion and politics and family, those are the three biggest things. Just women sitting around and talking would be fine with me. (Gina Billingsly)

Although not all middle-class black women registered acceptance, none of their statements about the need for change came close to these in the degree to which they envision the nature of the fundamental changes that would be necessary.

The greater acuteness of comments by working-class women in this subset of interviews cannot be taken as proof of Jaggar's assertions, only as an indication that they might well be true. Those at the bottom of this subset—black working-class women—suffer the "triple jeopardy" of race, class, and gender. This may increase the likelihood that they will rise to the necessity of achieving a critical stance, seeing the society as they do from its very margins and suffering discrimination on so many counts.

12 The Embodiment of Oppositions

If the Lord hadn't a knowed that the woman was gonna be stronger, he wouldn't a had the women have the periods and the babies. Can you see a man having a baby?
—Phyllis Hood

I began this study by revealing how women represent themselves as fragmented—lacking a sense of autonomy in the world and feeling carried along by forces beyond their control. It seems probable that the causes behind this fragmentation must be operating at a very general level, affecting women at widely separated ends of the social hierarchy. They probably include implicit scientific metaphors that assume that women's bodies are engaged in "production," with the separation this entails (given our conception of production) between laborer and laborer, laborer and product, laborer and labor, and manager and laborer. They probably also include causes outside the scope of this study, such as our tendency to see the self as needing for its completion the acquisition of various kinds of external commodities.[1]

I can only speculate why women's general images of their selves are chronically fragmented, but I would like to suggest (with somewhat more confidence) some reasons why the interviews contain—alongside fragmented images—such a rich mix of consciousness of alternative social and cultural worlds, together with resistance and protest against conditions perceived to be diminishing and denying of autonomy and fulfillment. Marcuse was struck by the "ever-growing number of people who, in a strict sense, cannot imagine a qualitatively different universe of discourse and action."[2] What is striking about our interviews with women is the extent of their questioning, opposing, resisting, rejecting, and reformulating the ways in which they live and the ways in which the society might work.

It may be that men and women question things to the same extent, but since I did not interview men about their everyday experiences, I cannot say. However, many studies of working-class consciousness find little evidence of resistance, and perhaps this is because those studies are focused on men.[3] Evidently many women can achieve a high degree of skepticism. What are some reasons this might be so?

In response to the problem of philosophical skepticism—how do we answer doubts about the justification of our beliefs?—Wittgenstein writes of the impossibility of trying to ground or prove much of what we believe to be true. We hear on authority, accept on authority, and confirm by our own experience a wide range of ideas that form the substratum of our thought about the world, the bedrock of any inquiries we might make about it. He has in mind things such as these: I have a body that does not disappear and reappear, the sun is not a hole in the vault of the heavens, there is a brain inside my head, I have two hands, I have grandparents, the earth exists, and so on. These things are taken-for-granted aspects of existence; we would not know what it would mean to call them into question.[4]

Some of Wittgenstein's examples are related to the practice of science, as in the epigraph of Chapter 3: experimenters assume a certain worldview before they begin to investigate, and their investigations do not call that worldview into question. The interesting anthropological question is how we draw the line between aspects of science that are part of the taken-for-granted nature of reality—the earth exists—and aspects that assume a view of the world that might well be different. An example of this would be the implicit assumption that body parts are organized in a bureaucratic hierarchy and subjected to central control. The problem of whether this is a taken-for-granted assumption or whether it can be questioned is relevant to this study if we suggest that aspects of science might reflect a worldview that is less comfortable to women than to men, and so more easily strikes women as questionable. Wittgenstein says somewhat glibly, "I learned an enormous amount and accepted it on human authority, and then I found some things confirmed or disconfirmed by my own experience."[5]

But what happens when men's experience and women's are distinctly different? One key to this question lies in the special status accorded to science as a form of knowledge. Basil Bernstein has distinguished two major linguistic codes, elaborated and restricted. Elaborated codes orient speakers toward such habits as describing events independently of their context, analyzing experience, or objectifying experience. Scientific language falls in this category. Restricted codes orient speakers toward such habits as describing events so they are dependent on their context, or taking preestablished meanings and values for granted. These codes differ in how they are used and emphasized by class: the working class emphasizes restricted codes, the middle class elaborated codes.[6] Dell Hymes has rephrased this distinction as a contrast between more implicit (restricted) and more explicit (elaborated) styles:

> It is not that one of the styles is "good," the other "bad." Each has its place. The more implicit style, in which many understandings can be taken for granted, is essential to efficient communication in some circumstances, and to ways of life in others. But, and this is an aspect of Bernstein's view that has often been overlooked, the more explicit style is associated with predominantly universalistic or context-free meanings, while the more implicit style is associated with particularistic or context-specific meanings. And, argues Bernstein, the universalistic meanings possible to the more explicit style are essential, if one is to be able to analyze means of communication themselves, the ways in which meanings come organized in a community in the service of particular interests and cultural hegemony, and so gain the knowledge and leverage necessary for the transformation of social relationships.[7]

Hymes has raised some extremely important criticisms of Bernstein's distinction, pointing out many implicit value judgments that Bernstein makes: since Bernstein argues that use of an elaborated code is necessary to analyze and so achieve transformation of the social order, it tends to be interpreted as more valuable, and its use more widely is interpreted as a good thing. Another implicit assumption is that science, held to be more abstract and context-free, is part of an elaborated code. Hence it is a desirable discourse in efforts to move toward human liberation.[8] Hymes asks whether science is really abstract and context-free or whether it is highly particular and context-dependent, and whether abstract discourse is necessarily the only or even the best way to achieve general understanding. He suggests that although the concrete, narrative, storytelling form of discourse is deprecated and drummed out of students at prestigious institutions, narrative forms are still used by the powerful: it is just that they use them in the ways they choose. Further, he gives us a glimpse of the subtle ways in which narrative can be used to "give a weighted quality to incident" and so achieve, through a restricted code, general understandings.[9] Stories of our concrete experiences, even anecdotes, can be insightful commentaries on the social order.

What I would like to suggest then is that the seemingly abstract code of medical science in fact tells a very concrete story, rooted in our particular form of social hierarchy and control. Usually we do not hear the story, we only hear the "facts," and this is part of what makes science so powerful. But women—whose bodily experience is denigrated and demolished by models implying failed production, waste, decay, and breakdown—have it literally within them to confront the story science tells with another story, based in their own experience.

We saw in Chapter 2 how numerous contrasts dominate postindustrial capitalist society: home versus work, sex versus money, love versus contract, nature versus culture, women versus men. Because of the nature of their bodies, women far more than men cannot help but confound these distinctions every day. For the majority of women, menstruation, pregnancy, and menopause cannot any longer be kept at home. Women interpenetrate what were never really separate realms. They literally embody the opposition, or contradiction, between the worlds. Sometimes the embodiment is private, as women menstruate through a day of work with no one else knowing or manage to conceal hot flashes, but of course the women themselves are acutely aware of what they conceal from others. Pregnancy, of course, cannot be concealed for long, and anyone who has looked for a job or continued one while pregnant cannot help but see the clash with which the two worlds, meant to be kept ideologically separate, collide.[10] A pregnant working woman is an embarrassment, an offense. She is threatened with loss of job or career, or it is assumed she will quit; she is told she never would have been hired if her supervisor had been warned, she is told she cannot have it both ways. The problem is that women have the potential for both ways in them all the time.

In addition, as I argued in Chapter 7, women find in the concrete experiences of their bodies a different notion of time that counters the way time is socially organized in our industrial society. In the universe of cultures, there are different ways of conceptualizing time that contrast with the one familiar to us, in which we measure it, treat it linearly, and think of it as something to be saved, bought, and sold. Benjamin Whorf described the Hopi notion of "burgeoning activity," in which what counts is the way a sequence unfolds (the growing of plants, formation of clouds, or building of a house), not the amount of time that elapses.[11] Others have documented how workers during the Industrial Revolution had to be taught new forms of organizing and regarding time.[12] Therefore it should not be inconceivable that women, grounded whether they like it or not in cyclical bodily experiences, live both the time of industrial society and another kind of time that is often incompatible with the first.

A clash between two fundamentally different concepts and experiences of time suggests a struggle, and there have been numerous struggles from the Industrial Revolution to our own day. In this book we have seen the struggles and debates around menstruation, PMS, birth, and menopause, the struggles for definitions of terms, of persons, of process, of meaning, and for the allocation of resources. The terms of the struggle have often been set by specifically political uses of science and the dominant cultural ideology: women cannot hold certain jobs because of their hormones, their strength, their size, their pregnancies;[13] if blacks have higher disease rates it must be because of genetic biological factors, not because they are exposed to greater stress and more carcinogens in the kinds of jobs they do.[14]

To combat these political uses of science and ideology, some feminist theorists have argued that there is a specifically feminist standpoint arising out of women's particular social and psychological construction of the self. Such feminists argue that because of men's very different life experiences (in the family and in the division of labor), they develop much more rigid boundaries around the self than women do. Based on object-relations theory, applied to women's social roles by Nancy Chodorow,[15] this line of thought sees men as constructing selves with a hostile and combative dualism (self-other) at their hearts.

> Masculinity must be attained by means of opposition to the concrete world of daily life, by escaping from contact with the female world of the household into the masculine world of public life. This experience of two worlds, one valuable, if abstract and deeply unattainable, the other useless and demeaning, if concrete and necessary, lies at the heart of a series of dualisms—abstract/concrete, mind/body, culture/nature, ideal/real, stasis/change. And these dualisms are overlaid by gender: only the first of each pair is associated with the male.[16]

Nancy Hartsock argues that in these contrasts men's experience inverts that of women, who, because of their deep involvement in the concrete processes of birth, childrearing, and housework, experience life as concrete, bodily, natural, real, and changing.

> The female construction of self in relation to others, leads . . . toward opposition to dualisms of any sort, valuation of concrete, everyday life,

> sense of a variety of connectednesses and continuities both with other persons and with the natural world. If material life structures consciousness, women's relationally defined existence, bodily experience of boundary challenges, and activity of transforming both physical objects and human beings must be expected to result in a world view to which dichotomies are foreign.[17]

In particular, women's experience confounds the dichotomy mental/manual labor.

> The unity of mental and manual labor, and the directly sensuous nature of much of women's work leads to a more profound unity of mental and manual labor, social and natural worlds, than is experienced by the male worker in capitalism. The unity grows from the fact that women's bodies, unlike men's, can be themselves instruments of production: in pregnancy, giving birth or lactation, arguments about a division of mental from manual labor are fundamentally foreign.[18]

As Dorothy Smith puts it, women work in "the bodily mode" and in so doing both allow others—mostly men—the freedom to work in the "abstracted conceptual mode" and conceal from them that what they do depends on the material details being taken care of by someone else (women).

> It is a condition of anyone's being able to enter, become and remain absorbed, in the conceptual mode of action that she does not need to focus her attention on her labors or on her bodily existence . . . If men are to participate fully in the abstract mode of action, they must be liberated from having to attend to their needs in the concrete and particular . . . Women . . . do those things which give concrete form to the conceptual activities. They do the clerical work, giving material form to the words or thoughts of the boss. They do the routine computer work, the interviewing for the survey, the nursing, the secretarial work. At almost every point women mediate for men the relation between the conceptual mode of action and the actual concrete forms on which it depends. Women's work is interposed between the abstracted modes and the local and particular actualities in which they are necessarily anchored. Also women's work conceals from men acting in the abstract mode just this anchorage.[19]

There is much in feminist theory that resonates with the empirical data of this study. However, one tenet of feminist theory does not find corroboration. As Alison Jaggar explains it, although there is a profoundly different view of the world encapsulated in women's experience,

> the standpoint of women is not expressed directly in women's naive and unreflective world view . . . women's male-dominated perceptions of reality are distorted both by male-dominant ideology and by the male-dominated structure of everyday life. The standpoint of women, therefore, is not something that can be discovered through a survey of women's existing beliefs and attitudes—although such a survey should identify certain commonalities that might be incorporated eventually into a systematic representation of the world from women's perspective. Instead, the standpoint of women is discovered through a collective process of political and scientific struggle.[20]

Of course, there are some ways in which the women in this study have their views distorted by the male-dominated structure of society. Most women feel shame at the prospect of soiling themselves publicly with menstrual blood and try earnestly to conceal it. The same goes for menopause. But we have seen that alongside this shame and embarrassment are a multitude of ways women assert an alternative view of their bodies, react against their accustomed social roles, reject denigrating scientific models, and in general struggle to achieve dignity and autonomy. The benefits of "a collective process of political and scientific struggle" are undeniable, but I would assert that everyday life is also a struggle and therefore can and does contain a critical standpoint, at least for some.

Because their bodily processes go with them everywhere, forcing them to juxtapose biology and culture, women glimpse every day a conception of another sort of social order. At the very least, since they do not fit into the ideal division of things (private, bodily processes belong at home), they are likely to see that the dominant ideology is partial: it does not capture their experience. It is also likely that they will see the inextricable way our cultural categories are related and so see the falseness of the dichotomies. When women derive their view of experience from their bodily processes as they occur in society, they are not saying "back to nature" in any way. They are saying on to another kind of culture, one in which our current rigid separations and oppositions are not present.

However, the existence of awareness and questioning of the status quo does not necessarily lead to revolutionary action or social change. "Indeed they could conceivably be the grounds for inactivity and a debilitating pessimism. A heightened consciousness of the limits and bounds of one's oppression may result in a surrender to the awesome power of one's oppressors."[21] But although this study provides no way to estimate the chances of successful social change in gender inequality, at least it makes clear that women not only are often aware of their oppression, but they are able to forge alternative visions of what the world might be like. Looked at as a program for social reform, women's responses to science and ideology are not trivial. They are radical, they are threatening, they would mean revolution.

This leads me to a final hunch. I did not know whether I would have grounds at the end of this study for saying that working-class women express at least as much consciousness and resistance as middle-class women. I now feel confident that this is so. Why should this be? If we think simply for a moment, what all women in our society share is the experience of the housekeeping of their own bodies, whose effluvia, demands, and exigencies so seldom appear in schedule with socially organized time. In addition, almost all women share the primary responsibility (if not the actual job) of their family's housekeeping—cleaning dirty floors, diapers, and toilets, taking out the garbage, washing clothes. We have seen how some feminist theorists argue that because women engage in this kind of work they are more grounded than men in concrete activity and practical experience and more able to see the social whole, which includes both its concrete and its abstract parts.

If there is anything at all to the relationship between housekeeping of the "body" of the family—of its effluvia, dirt, waste—and a different and more practical, grounded consciousness prone to question the shape of society as a whole, then there should be a more acute consciousness in the working class. For they, especially women, and most especially black women, in large part do the housekeeping for the whole social "body" in addition to the housekeeping for their own bodies and their families. It is they who clean rich people's homes, clean offices and factories, take away the waste and garbage of cities and towns, serve people at restaurants and clean up after them, care for the daily needs of patients and clean up after them in hospitals.[22]

We must not make the mistake of hearing the particularistic, concrete stories of these and other women and assume that they are less likely than more universalistic, abstract discourse to contain an analysis of society. It is up to anyone who listens to a woman's tale to hear the implicit message, interpret the powerful rage, and watch for ways in which the narrative form gives "a weighted quality to incident," extending the meaning of an incident beyond itself.

The concrete incidents of women's everyday lives can evoke glimpses of other ways of living, other ways of using time, other ways of conveying the sense of menstruation, birth, and menopause. Until hearing these words, the best way I knew to evoke alternative visions was to look to poetry, such as this:

Vision begins to happen in such a life
as if a woman quietly walked away
from the argument and jargon in a room
and sitting down in the kitchen, began turning in her lap
bits of yarn, calico and velvet scraps,
laying them out absently on the scrubbed boards
in the lamplight, with small rainbow-colored shells
sent in cotton-wool from somewhere far away,
and skeins of milkweed from the nearest meadow—
original domestic silk, the finest findings—
and the darkblue petal of the petunia,
and the dry darkbrown lace of seaweed;
not forgotten either, the shed silver
whisker of the cat,
the spiral of paper-wasp-nest curling
beside the finch's yellow feather.
Such a composition has nothing to do with eternity,
the striving for greatness, brilliance—
only with the musing of a mind
one with her body, experienced fingers quietly pushing
dark against bright, silk against roughness,
pulling the tenets of a life together
with no mere will to mastery,
only care for the many-lived, unending
forms in which she finds herself,
becoming now the sherd of broken glass
slicing light in a corner, dangerous
to flesh, now the plentiful, soft leaf
that wrapped round the throbbing finger, soothes the wound;
and now the stone foundation, rockshelf further
forming underneath everything that grows.[23]

One danger in listening either to poetry or to ordinary women's talk is that we will romanticize women's special ability to see the truth about life, basing this ability on a kind of essentialist, natural proclivity that only women have. My claim has been rather that those at the bottom of the heap tend to see more deeply and clearly the nature of the oppressions exacted by those at the top of the heap. In our society those at the bottom are often female and black, but if this were somehow reversed, I would expect white males to gain relatively greater critical vision.

Whenever a researcher gathers together the talk of many individuals, organizes it around what appear to be common themes, and presents these back to those who talked in the first place, the result is hard to predict. In her study of a working-class neighborhood Lillian Rubin found such intense frustration, sorrow, and hardship that those she interviewed felt these collective truths were difficult to accept and met them defensively.[24] When the streams of talk we collected are gathered together, many hard truths are also revealed. But in addition, putting together many individual voices has produced a resounding chorus. The exhilaration and the wisdom in this chorus tell us of many visions of life, different for different women and powerfully different from the reality that now holds sway.

Appendix 1: Interview Questions

These are the questions we used as a guide in the interviews, grouped roughly according to topic.

Background

What is your age? What are the occupations of your mother, your father, yourself, your partner or spouse if any, your children if any?

Who lives in your household or family now? Who lived in the family you grew up in?

What is done in your family for the sake of health (diet, exercise, etc.)? What is done for the sake of illness?

Menstruation

How did your family regard menstruation? What did you know about it before your first period, from your parents, siblings, friends, books, or school?

What was it like the first time you menstruated?

How did you feel about menstruating during your early years?

Were there any special practices or restrictions at school, work, or home about diet, exercise, bathing, or anything else during your period?

What significance in your life did beginning to menstruate have?

How would you explain menstruation to a young girl who knew nothing about it? How did you explain it to your daughters?

Do you feel different before, during, or after your period?

Do you do anything special or refrain from doing anything around the time of your period?

Does your period affect you at school, work, athletics, etc.?

What is your (nonscientific) understanding of why menstruation occurs?

Does it have any other (religious, spiritual) significance for you?

Overall how do you feel about menstruating? About the prospect of menstruating for the next thirty (or however many) years?

How would you react if someone magically offered you the chance never to menstruate again?

Do you remember your first gynecological exam? What was it like for you?

Can you suggest any changes in society that would make it easier to be a woman, to be a menstruating woman?

Childbearing

If currently pregnant

What have you done differently (if anything) during the pregnancy, such as diet, exercise, sleep, clothing?

How have you been feeling (physically, emotionally)?

Has pregnancy affected your activities (paid work, housework, school, etc.)?

What medical prenatal care have you had (checkups, sonogram, genetic tests, other tests of any sort)? How did you feel about any of them?

Looking forward to labor, what thoughts or expectations do you have? What have you heard about it from mother, sisters, friends? If you could have it go any way you wanted, how would it go? What would you like the role of doctors, nurses, etc., to be? What would you like the role of friends or family to be? How do you feel about medicine to make the labor start or medicine to make the labor go faster? How do you feel about anesthetic to reduce your discomfort? How would you respond if your doctor recommended any of these? What would happen if you disagreed?

Do you want rooming-in? Do you plan to breast-feed? Any special thoughts about either?

What plans do you have for the baby afterward, in terms of going back to work, looking for work, child care, taking care of the baby, etc.?

If you could arrange things any way you liked, how would they be?

Past pregnancies

How did you feel during earlier pregnancies compared with this one?

What were the circumstances and experiences of each pregnancy, labor, and delivery? Was anything special done for your health during pregnancy? (Any prenatal care, any tests or procedures?) How did you feel during each pregnancy? What was labor like? (When did it start, how long did it last, how did the different parts of labor feel, when did you go to the hospital, and who decided this?) Any medication? (Induction, medication to speed up or slow down contractions, any analgesia or anesthesia, any monitoring of fetus or you? IV? Forceps? Cesarean section?) How were the decisions made for or against these? How did you feel about this?

How soon after the birth did you see your baby? Have care of him or her? Was this all right with you? Is there anything you wish had been different about this?

What was the involvement of parents, husband, doctors or midwives, hospital staff, other children, friends?

Can you remember any particular hopes or fears you had around the time of your children's births?

Is there anything about the births of your children that you wish could have been different?

If never pregnant

Looking ahead, do you have any hopes or fears about pregnancy, about labor?

What have you heard or learned about it from your parents, siblings, friends, books?

If you were to get pregnant, what kind of birth experience would you want?

Menopause

Younger women

Have any of your older relatives or friends experienced menopause? What was it like for them? What were they like during it?

What is your understanding of why menopause occurs?

Looking ahead, do you have any hopes and fears about your own menopause?

Older women

Have you experienced menopause (change of life, etc.)? How did you know it was happening? Have you done anything for it/about it? Have doctors, friends, mother, children been involved? What is your understanding of why women experience this change? How do you feel about it? Does it have any special meaning for you, positive or negative?

Have you ever experienced a hot flash? Can you describe what it felt like? Can you remember the circumstances of any particular one? What happened before and after?

Were you embarrassed about it? Did others notice, and were they embarrassed? If so, why do you think hot flashes are embarrassing to people?

How would you explain menopause to a young woman?

General

Do you think there is anything that would make women's lives in general easier or better than they are?

What did you think of the interview?

Appendix 2: Biographical Profiles

The total of 165 interviews consisted overall of 29% with women in the youngest group (puberty to childbearing), 42% with women in the middle group (childbearing and childrearing), and 29% with women in the oldest group (menopause and beyond). 43% of the interviews were with working-class women and 57% with middle-class women. 28% were with black women, 72% with white women. The chart shows the numerical distribution.

Life Stage	youngest		middle		eldest	
Working class						
	a		**c**		**e**	
	black	11	black	17	black	8
	white	5	white	19	white	11
		16		36		19
Middle class						
	b		**d**		**f**	
	black	6	black	4	black	1
	white	26	white	30	white	27
		32		34		28

In the biographical sketches that follow, the letter at the end (keyed to the chart) indicates the class and life stage in which I determined that each woman fell. The definitions of class I used as a guide are on pp. 5–6. All names are pseudonyms and occupations are stated generally enough to protect people's identity.

Ansell, Linda
14 years old. She is a high school student, living with her father, a financial analyst. (b)

Anthony, Theresa
In her twenties. Before this pregnancy, she was a retail manager for a department store; her husband is in business for himself. She is seven months pregnant with her first child. (d)

Appell, Claudia
30 years old. She is a nurse; her husband is a salaried employee of a large company. She has two children. (d)

Baker, Carmelita
15 years old, black. She is a high school student. Her mother is a community worker; her father is a minister's assistant. (a)

Barrows, Ginny
31 years old. She is on welfare, separated from her husband and living with her three young children. (c)

Barton, Patricia
21 years old. She is on welfare, living with her father, a machine operator. She has a four-year-old child. (c)

Bartson, Valerie
16 years old, black. She is a high school student; her mother is a part-time tutor, separated from her father. (a)

Benson, Ella
In her twenties, black. She is a physical education teacher, unmarried. (b)

Bergerson, Carol
34 years old. She is a housewife; her husband teaches at a high school. She has three children, all born by cesarean section. (d)

Berg, Alma
25 years old. She does clerical work at a construction company; her husband is a truck driver. She is three months pregnant with her first child. (c)

Berger, Wendy
In her thirties. She is a teacher; her husband is a businessman. She has two children. (d)

Berkeley, Susan
In her mid-twenties. She is not working because of injuries. Her husband is an engineer. She is seven months pregnant with her first child. (d)

Billingsley, Gina
25 years old, black. She is a security agent and domestic worker; her husband is a truck driver's assistant. She has two children. (c)

Brown, Lucille
68 years old, black. She is a retired clerical worker; her husband was a longshoreman before his death. (e)

Bucci, Margaret
In her late twenties. Before her children were born she was an abortion clinic worker; her husband is a reporter. She has two children, both by cesarean section. (d)

Bunker, Linda
18 years old, black. She is in high school. Her mother works in the food service industry and is separated from her father. (a)

Burks, Hattie
65 years old, black. She is a retired domestic worker; her husband was a factory worker. She has three children. (e)

Carlson, Ruth
52 years old. She is a librarian and has never been married. (e)

Carson, Lois
58 years old, black. She is retired; formerly she worked as a domestic. Widowed, she has three grown children. (e)

Chapman, Ruth
65 years old, black. She is a retired domestic worker. Her husband was a school bus driver before his death. (e)

Cook, Juliet
17 years old, black. She has plans to finish high school at a vocational school. She is unemployed and is raising her three-year-old child. (c)

Corbin, Carrie
23 years old, black. She is a student at a state college, living with her mother, a homemaker, and her father, a minister. (b)

Craft, Marge
In her late forties. She is a homemaker; her husband owns his own business. She has four children. (f)

Crawford, Helen
84 years old. She is a retired secretary; her husband was a hotel manager before he died. (f)

Cresswell, Teresa
31 years old. She works for her husband, a self-employed plumber. She has four children. (d)

Crichton, Margaret
32 years old, black. She owns her own business, is widowed, and is raising her five children. (d)

Cromwell, Laura
In her thirties. She is a housewife; her husband is a scientist. She has two children, the first by cesarean section. (d)

Crowder, Ellen
In her mid-twenties. She is a research assistant in a scientific lab; her husband is a postdoctorate researcher at a university. Her newborn baby, born by cesarean section, is her first. (d)

Davis, Adrian
18 years old, black. She is a high school graduate, between jobs. She lives with her parents, both teachers, and her six-week-old child. (d)

Dinton, Cheryl
23 years old, black. She formerly worked as a teacher's aide. She has a five-year-old child and is seven months pregnant. (c)

Dorset, Lucy
In her late forties. She is a full-time volunteer; her husband is an attorney. She has two children. (f)

Doyle, Mary
In her late forties. She is a special educator; her husband is a psychiatrist. She has three teenage children. (f)

Duke, Anita
24 years old. She is a part-time receptionist and lives with her two young children. (c)

Elger, Minnie
50 years old. She babysits children in her home. Separated from her husband, she has two children. (e)

Evans, Gracie
53 years old, black. She works in a food store, has seven children, and is divorced. (e)

Fitsik, Emma
19 years old. She is attending college. Her mother is a homemaker; her father is a teacher. (b)

Frank, Sylvia
58 years old. She is a retired mental health worker, divorced. She has two children. (f)

Fried, Liza
In her late forties. She is a psychiatric social worker; her husband is a doctor. She has two children. (f)

Garcia, Joann
20 years old. She is a social worker at a nursing home for the elderly; her husband is a technician at a hospital. She is eight months pregnant with her first child. (d)

Gibson, Martha
63 years old. She is a retired secretary and bookkeeper. Never married, she lived for many years with her mother, a housewife, and her father, a banker. (f)

Gleason, Carol
In her mid-twenties. She works for a community organization; her husband is head of a nonprofit organization. She has one child. (d)

Gonzalez, Chris
A nurse in her late twenties. Her husband is a doctor. They have one child, born a few months before the interview. (d)

Green, Rita
66 years old. She is a retired high school teacher divorced from her husband. She has three children. (f)

Griffin, Joyce
33 years old, black. She lives with her husband (unemployed), her four children, and two grandchildren. (c)

Gruner, Ann
In her early thirties. She is a clerical worker; her husband is a skilled craft worker. She is four months pregnant with her first baby. (c)

Guido, Dorothy
36 years old. She is unemployed; formerly she worked as a waitress. She has four children. (c)

Hammond, Sandy
16 years old, black. She is a high school student, living with her mother, a hospital worker. Her parents are separated. (a)

Heath, Barbara

70 years old. She is a retired dietician who has never married. She lived with her mother, a homemaker, and her father, a restaurant manager, until they died. (f)

Henderson, Patricia

16 years old, black. She is in high school; her mother is a nurse's aide, separated from her father. (a)

Hernandez, Rafaela

20 years old. She is in college: her mother is a beautician; her father is an unemployed salesman. (a)

Herrick, Jean

33 years old. She works as a data processor and has never been married. (b)

Hessler, Carla

20 years old. She is a senior in college: her mother is a homemaker; her father is retired from a military career. (b)

Hobbs, April

28 years old, black. She is a nurse's aide. She lives with her five sisters, her mother, and her four-week-old daughter. (c)

Hoffman, Estelle

61 years old. She is a supervisor in a government agency; her husband is a dentist. She has two children. (f)

Hood, Phyllis

In her sixties. She is a homemaker. Formerly she did administrative work for her husband's small business. She has two children. (f)

Hooper, Regina

In her seventies, black. She formerly worked on an assembly line. She went to college at age 47 and got a degree, but she has been unable to find a job. She has two children.

Hubbard, Sarah

63 years old. She is a retired secretary, widowed. She has two adult sons. (e)

Hunt, Barbara

22 years old, black. She is a student at a secretarial school. She has two children and is seven months pregnant. (c)

Hunter, Pamela

32 years old. She is a housewife; her husband is a construction worker. She has two children, both born by cesarean section. (c)

Hurley, May
63 years old, black. She has had a wide variety of paid and volunteer jobs from a pharmacy driver to a hospital worker. She has nine living children. (e)

Hurst, Sharon
38 years old. Before her children were born she was a nurse administrator; she is not working now. Her husband is a research scientist. She has two small children. (d)

Hutchins, Zena
25 years old, black. She was a security guard until she got pregnant. Now she lives with her ten-month-old child, her mother, a nurse, and her stepfather, who is retired. (c)

Jackson, Sue
In her thirties. She is a housewife; her husband is a carpenter. She has two children, the first born by cesarean section. (d)

Jacobson, Ethel
70 years old, black. She is a retired garment trade worker, divorced with one grown child. (e)

Jankowski, Nancy
In her thirties. She was a government worker before having children; her husband is a lawyer. She has three children, the first born by cesarean section. (d)

Jefferson, Molly
34 years old. She was a teacher before her children were born; her husband is a businessman. She has two children, the first born by cesarean section. (d)

Johnston, Marjorie
In her late forties. She is a program developer for a private school. Her husband is a CPA. She has three children. (f)

Jones, Bernadette
27 years old, black. She is a part-time nurse's assistant; her husband is a mechanic. She has four children. (c)

Kelley, Agnes
81 years old. She worked in a factory and as a farm laborer while raising two children. Widowed, she now lives alone. (e)

Kelsey, Jody
16 years old, black. She is in high school. Her mother is a psychiatric nurse's aide, separated from her father. (a)

Kidwell, Mildred
56 years old. She is a full-time secretary; her husband is retired. She has two children. (e)

King, Mary
17 years old, black. She is a senior in high school. Her mother does clerical work; her father is a tractor-trailer driver. (a)

Kirschner, Judy
A 21-year-old college student who earns extra money as a babysitter. She is pregnant with her first child. The baby's father, with whom she lives, has been accepted into medical school. (d)

Koerner, Shirley
In her early twenties. She is a part-time health educator in a hospital. Her husband is a Ph.D. candidate. She is five months pregnant with her first child. (d)

Kuper, Dolores
69 years old. She is retired from a professional career and widowed. She has two children. (f)

Ladd, Pat
32 years old. She teaches photography at a community college; her husband is a businessman. She is two months pregnant with her first child. (d)

Larkin, Terry
In her late twenties. She is a proofreader for a local press; her husband is a researcher in a private laboratory. She is eight months pregnant with her first child. (d)

Larrick, Alice
19 years old. She is a high school senior who lives with her boyfriend. Her mother is self-employed and is divorced from her father. (b)

Larson, Elizabeth
31 years old. She worked as a maid until she was several months pregnant. Unmarried, she has returned to live with her mother. She is eight months pregnant with her first child. (c)

Lasch, Sarah
In her twenties. She teaches ballet and childbirth education classes; her husband is a businessman. She has two children, the first born by cesarean section. (d)

Lasley, Mary Jo
17 years old. She lives with her mother, a homemaker, and her father, a mechanic. She has a fourteen-month-old child. (c)

Lassiter, Kristin
14 years old. She is a high school student. Her mother does university research; her father is an actor. (b)

Latham, Juliet
18 years old, black. She is studying for her high school equivalency diploma and is five months pregnant with her first child. (c)

Lehman, Rachel
20 years old. She is a college senior. Her mother is a psychotherapist; her father is a computer systems analyst. (b)

Lenhart, Mara
20 years old. She is a college student: her mother is a writer, her father an architect. They are divorced. (b)

Lentz, Rebecca
In her mid-twenties. She taught special education before getting pregnant. Her husband is a chemist. She is two months pregnant with her first child. (d)

Levinson, Shelley
21 years old. She is a university student. Her mother is a home economist, separated from her father. (b)

Levy, Marcia
29 years old. Before this pregnancy she was a buyer for a store. Her husband is a salesman. She is nine months pregnant with her second child. (d)

Litson, Jean
In her late forties. She is an administrative assistant for a religious organization; her husband is in business. She has one child. (f)

McGuire, Robin
In her early twenties. She worked as a waitress before getting pregnant with her second child eight months before the interview. Her husband is a mechanic. (c)

Madsen, Kathy
In her early twenties. She is a nurse's aide; her husband is an auto mechanic. Her newborn baby is her first. (c)

Mann, Betsy
In her mid-twenties. Her husband is a salaried businessman. She

does not wish to work outside the home. She is eight months pregnant with her first child. (d)

Marriott, Rhonda
15 years old. She is a tenth-grade student at a public school. Her mother is a housewife, her father an engineer. (b)

Matthews, Linda
24 years old, black. She was a state office worker and is now a student at a community college. She has a five-year-old child born by cesarean section and is seven months pregnant. (c)

Miller, Eileen
35 years old, black. She is a pentacostal pastor; her husband is a maintenance worker. She has two teenage children. (c)

Miner, Lisa
18 years old. She is a high school graduate, now working at a food store, planning to go to college next year. She lives with her father, a professor, and her mother, a graduate student. (b)

Minton, Elsie
In her sixties, black. She is a retired federal employee. (f)

Mitchell, Vivian
25 years old. She is an office worker in a computer firm; her husband is a skilled craft worker. She is eight months pregnant and has one other child. (c)

Monroe, Dorothy
55 years old. She has been a teacher in public school for thirty years. (f)

Moore, Jan
In her late twenties. She is a health worker; her husband is a bus driver. She is eight months pregnant with her first child. (c)

Morgan, Julie
16 years old. She is a senior in high school, living with her mother, a teacher, and her father, a state employee. (b)

Morrison, Ann
37 years old. She is a secretary and part-time student. She has a three-year-old child. (c)

Morse, Kelley
17 years old, black. She is a high school student. Her mother is a clerical worker in a hospital, separated from her father. (a)

Mullendorf, Doris
In her fifties. She teaches part-time for the federal government; her husband is in publishing. She has two children. (f)

Naylor, Stephanie
26 years old, black. She is a clerk at a hospital. She lives with her mother (who is disabled) and her 11-year-old child. (c)

Neilson, Pat
59 years old. She is a federal employee; her husband is a businessman. (f)

Nessler, Michelle
23 years old, black. She is a counselor at a state college and lives with her fiancé. (b)

Nichols, Teddie
19 years old, black. She dropped out of high school. She lives with her boyfriend and two children. (c)

Norton, Edna
65 years old. She is a part-time volunteer; her husband is disabled and unemployed. She has four children. (e)

Nugent, Pamela
21 years old. She was training to be a nurse until her child was born. The father of her nine-week-old baby is a graduate student, with whom she lives. (d)

O'Hara, Meg
28 years old. She does data processing and formerly worked as a music teacher. She has not married. (b)

Oliver, Rose
15 years old. She is trying to get her high school equivalency diploma. She lives with her father, a skilled craft worker. (a)

Olmstead, Jeanine
17 years old. She dropped out of school and now does part-time babysitting. Her mother works at a local store; her father is a skilled craft worker. (a)

Olson, Laura
21 years old. She is a college senior. Her mother is a homemaker, her father a military officer. (b)

Ostrov, Martina
61 years old. She is a state employee. Widowed, she has three grown children. (f)

Parrish, Tania
20 years old. She is a university student. Her mother works at a school; her father owns a manufacturing company. (b)

Parsons, Nancy
21 years old, black. She is a junior at a state college. Her mother is a librarian, her father a teacher. (b)

Payne, Amanda
18 years old. She is a college student. Her mother is a landscape architect, her father is president of a company. (b)

Perdoni, Anna
19 years old. She is a college sophomore. Her mother is a homemaker, her father a dentist. (b)

Peterson, Janice
24 years old, black. Before this pregnancy, she worked as a bank teller. She has a four-year-old child and is pregnant with a second child. (c)

Phillips, Abbie
20 years old. She is a university student. Her mother is a teacher, her father a businessman. (b)

Pittman, Eleanor
72 years old. She is an unmarried retired health worker. (e)

Polasky, Gladys
70 years old. She is a retired telephone company typist, now widowed. (e)

Potter, Carolyn
20 years old. She is a university student. Her mother is a professor, divorced from her father. (b)

Powers, Jenny
In her early twenties. She is a clerical worker for an insurance company; her husband drives an ice cream truck. She is four months pregnant with her first child. (c)

Randals, Vivian
In her late forties. She administers a small public-interest organization. Divorced, she has two children. (f)

Reardon, Kathleen
18 years old, black. She is a junior in a public high school. Her mother is a clerical worker, her father a factory worker. (a)

Roark, Cathy
17 years old. She is trying to get her high school equivalency diploma while working part time as a sales clerk. Her mother works as a sales clerk; her father is a skilled craftsman. (a)

Robbins, Marcia
19 years old. She is studying for her high school equivalency diploma. Her mother is a shop clerk. She has a three-year-old child and is separated from the child's father. (c)

Ross, Linda
19 years old, black. She is a junior at a state college. Her mother does home care for the elderly; her father is an assistant manager of a store. (b)

Rubenstein, Leah
20 years old. She is a college senior. Her mother is a state employee, her father a publisher. (b)

Russell, Lois
In her late forties, black. She is a secretary; her husband is a machinist. She has two grown children. (e)

Sanchez, Mary
In her twenties. She is a physical education teacher; her husband is a graduate student. She has two children. (d)

Sanderson, Janice
34 years old. She worked as a secretary before her children were born. Her husband is a skilled craft worker. She has two children, the first born by cesarean section. (c)

Saunders, Margaret
In her late teens. She worked in a fast-food restaurant before getting pregnant. Unmarried, she returned home to live with her parents. Her father is a carpenter. She is eight months pregnant with her first child. (c)

Schiffer, Marion
50 years old. She is a social worker; her husband is a psychiatrist. She has four grown children. (f)

Scott, Crystal
18 years old, black. She is studying to get her high school equivalency diploma and has one infant. (c)

Simmons, Marguerite
34 years old, black. She is a health clinic coordinator. She is separated from her husband and lives with her two children. (d)

Sokolov, Becky
20 years old. She is on welfare while living with her mother. Her newborn child is her first. (c)

Spencer, Kathy
15 years old, black. She is in high school; her mother works in the service industry and is separated from her father. (a)

Stephens, Lisa
16 years old, black. She is in high school; her mother is a nurse's aide, separated from her father. (a)

Stetson, Joycelyn
In her late forties. She is a homemaker; her husband is a federal employee. She has two children. (f)

Stompers, Lisa
19 years old. She attends a university. Because of health problems, she lives with her mother, who is an information specialist. (b)

Summerdale, Edna
72 years old. She is a retired librarian, widowed. She has two children. (f)

Sundquist, Gladys
64 years old. She is a retired secretary; her husband is a retired carpenter. (e)

Taylor, Janie
66 years old. She now works as a volunteer with disabled people. When she was younger she worked as an agricultural laborer and in a factory. She is a widow with one adult child. (e)

Thomson, Georgia
18 years old, black. She just graduated from high school, is unemployed, and is raising her four-year-old child. (c)

Turnbull, Barbara
28 years old, black. She is a librarian at a public library, living with her mother, a federal employee. (b)

Turner, Naomi
In her sixties. She works part time at a health clinic; her husband is a retired businessman. She has two children. (f)

Tyson, Joan
In her late twenties. She is a housewife; her husband is a businessman. She has two children, the first born by cesarean section. (d)

Von Hausen, Freda
70 years old. She is a professor at a state college; her husband is retired. She has one child. (f)

Warren, Rebecca
38 years old, black. She works in administration at a state college. She is divorced and lives with her two teenage children. (d)

Wassen, June
57 years old. She is a telephone operator; her husband is a driver for a firm. She has two children. (e)

West, Alicia
19 years old. She is a student in a small college. Her mother is an engineer, her father a minister. (b)

Westover, Karen
18 years old. She is a college student who went to private secondary school. Her mother is an accountant, her father president of a company. (b)

Wilcox, Ruth
50 years old. She is a retired administrative assistant. Her husband is a car dealer. (f)

Williams, Claudia
65 years old. She is a retired nurse; her husband, from whom she is divorced, worked for an oil company. She has seven children. (f)

Windell, Eileen
25 years old, black. She is a financial counselor at a community clinic; she lives with her mother and her four-year-old child. (c)

Windsor, Ella
21 years old. She works at a food store; her mother is in university administration; her father is a newspaper editor. (b)

Wood, Mary
In her early twenties. She works with retarded children. Her husband is a salaried businessman. She is seven months pregnant with her first child. (d)

Wright, Phyllis
32 years old. She worked as a manufacturer's representative before she became pregnant. Her husband is a manufacturer's representative. She is seven months pregnant with her first child. (d)

Xenos, Sally
In her mid-twenties. She is a teacher at a private secondary school

working on her master's degree. Her husband is a salaried government worker also enrolled in a higher degree program. She is four months pregnant with her first child. (d)

Yager, Gerry

15 years old. She lives with her mother, who is self-employed. (b)

Yamada, Ellie

19 years old. She is a college student; her mother is a dietician, her father a dentist. (b)

Notes

Chapter 1: The Familiar and the Exotic

1. Taussig 1980:6
2. Strathern 1984:13
3. Geertz 1983:157
4. Benenson 1980:115
5. Rapp 1982
6. Benenson 1980:115; Robert Rothman 1978: 160
7. Robert Rothman 1978:162; Braverman 1974
8. Sennett and Cobb 1972
9. Rapp 1982
10. Quotations from interviews have been edited as lightly as possible to remove redundancies. Questions asked by the interviewer and brief clarifications are in brackets.
11. Marx 1967b:779. I am indebted to Ashraf Ghani for this reference.
12. Lewontin et al. 1984:10
13. Lewontin et al. 1984:11
14. Lewontin et al. 1984:11
15. Berliner 1975; Brown 1979; Burrow 1977
16. Friedson 1970; Segal 1984
17. Lock 1982

Chapter 2: Fragmentation and Gender

1. See Elshtain 1981 for the various forms the public-private split has taken since Greek civilization.

2. Zaretsky 1976; Sheila Rothman 1978
3. Tilly and Scott 1978:228
4. Fee 1976:176
5. Strathern 1984
6. Jaggar 1983:321–22
7. Schneider 1969:119
8. Barnett and Silverman 1979:42
9. Strathern 1984:17
10. Taub and Schneider 1982
11. Berreman 1972. See Brittan and Maynard 1984 for a theoretical analysis of racism and sexism and Gilman 1985 for a historical one.
12. Ollman 1976:133–35
13. Aronowitz 1973; Blauner 1964; Braverman 1974; Finifter 1972; Israel 1971; Seeman 1975; Shepard 1971, 1977
14. Ollman 1976:131–32
15. Sennett and Cobb 1972:208
16. Jaggar 1983:310
17. Jaggar 1983:311, 312
18. Jaggar 1983:315
19. Venable 1945:50
20. Frank 1981
21. House of Representatives 1983
22. Henderson 1985
23. Henderson 1985
24. Corea 1985
25. Foucault 1979:10, 11
26. Bartky 1982:138
27. Bartky 1982:134, 138
28. Keller 1985:89
29. Jaggar 1983:316
30. Fee 1981:381
31. Ardener 1978
32. Boggs quoted in Greer 1982:305

Chapter 3: Medical Metaphors of Women's Bodies: Menstruation and Menopause

1. Laqueur 1986:3, 18–19
2. Laqueur 1986:10
3. Rosenberg 1979:5
4. Rosenberg 1979:5–6
5. Tilt 1857:54
6. Tilt 1857:54, 57
7. Crawford 1981:50

8. Crawford 1981:50
9. Rothstein 1972:45–49
10. Crawford 1981:63
11. See Luker 1984:18; Crawford 1981:53–54; Skultans 1985
12. Laqueur 1986:4
13. Laqueur 1986:8
14. Fee 1976:180
15. Bagehot, quoted in Fee 1976:190
16. Geddes 1890:122
17. Geddes 1890:123
18. Geddes 1890:270–71
19. Barker-Benfield 1976:195–96
20. Sontag 1977:61–62
21. Geddes 1890:244; see also Smith-Rosenberg 1974:28–29
22. Smith-Rosenberg 1974:25–27
23. Quoted in Laqueur 1986:32
24. Ellis 1904:284, 293, quoted in Laqueur 1986:32
25. Smith-Rosenberg 1974:30–31; Wilbush 1981:5
26. Currier 1897:25–26
27. Haber 1983:69. See Good 1843:23–25 for an explanation of why the climacteric affects men more severely than women.
28. Smith-Rosenberg 1974:30
29. Taylor 1904:413
30. Quoted in Berliner 1982:170–71
31. Lewontin et al. 1984:58
32. Lewontin et al. 1984:59; see also Guyton 1986:23–24
33. In an extremely important series of papers, Donna Haraway 1978, 1979 has traced the replacement of organic functional views in biology by cybernetic systems views and shown the permeation of genetics and population biology by metaphors of investment, quality control, and maximization of profit.
34. Lewontin et al. 1984:59
35. Horrobin 1973:7–8. See also Guyton 1984:7. In general, more sophisticated advanced texts such as Guyton 1986:879 give more attention to feedback loops.
36. Lein 1979:14
37. Guyton 1986:885
38. Benson 1982:129
39. Netter 1965:115
40. Norris 1984:6
41. Dalton and Greene 1983:6
42. Mountcastle 1980:1615
43. Guyton 1986:885
44. Guyton 1986:885

45. Lein 1979:84
46. Evelyn Fox Keller 1985:154–56 documents the pervasiveness of hierarchical models at the cellular level.
47. Norris 1984:6
48. Giddens 1975:185
49. McCrea 1983
50. Lein 1979:79, 97
51. O'Neill 1982:11
52. Vander et al. 1985:597
53. Norris 1983:181
54. Vander et al. 1985:598
55. Norris 1983:181
56. Netter 1965:121
57. Netter 1965:116
58. Vander et al. 1985:580
59. Guyton 1986:968
60. Vander et al. 1985:576
61. Lein 1979:43
62. Guyton 1986:976
63. Vander et al. 1985:577
64. Guyton 1986:976; see very similar accounts in Lein 1979:69, Mountcastle 1980:1612, Mason 1983:518, Benson 1982:128–29.
65. Ganong 1985:63
66. Winner 1977:185, 187
67. Ewbank 1855:21–22
68. Ewbank 1855:23
69. Ewbank 1855:27
70. Ewbank 1855:141; on Ewbank, see Kasson 1976:148–51.
71. Fisher 1967:153
72. Fisher 1967:153; 1966
73. Mumford 1967:282
74. Mumford 1970:180
75. Noble 1984:312
76. Bullough 1975:298
77. Mason 1983:525
78. Guyton 1984:624
79. Vander et al. 1980:483–84. The latest edition of this text has removed the first of these sentences, but kept the second (Vander et al. 1985:557).
80. Vander et al. 1985:577
81. Vander et al. 1985:567, 568
82. Mason 1983:419; Vander et al. 1985:483
83. Ganong 1986:776
84. Mason 1983:423
85. Guyton 1984:498–99

86. Sernka and Jacobson 1983:7
87. Vander et al. 1985:557–58; Ganong 1985:356
88. Novak 1944:536; Novak and Novak 1952:600
89. Novak et al. 1965:642
90. See McCrea and Markle 1984 for the very different clinical treatment for this lack in the United States and the United Kingdom.
91. Kaufert and Gilbert 1986:8–9; World Health Organization Scientific Group 1981
92. Guyton 1986:979
93. Jones and Jones 1981:799
94. Sadly enough, even the women's health movement literature contains the same negative view of menstruation—failed production—as does scientific medicine. See Boston Women's Health Book Collective 1984:217 and Federation of Feminist Women's Health Centers 1981:74. As in the case of prepared childbirth literature, this is evidence of the invisible power of the ideology of the dominant culture.

Chapter 4: Medical Metaphors of Women's Bodies: Birth

1. Wertz and Wertz 1977:32. Merchant 1980 and Westfall 1977 provide a rich understanding of mechanical metaphors as they have been used to describe nature and humans since the seventeenth century.
2. Wertz and Wertz 1977:54–59
3. Wertz and Wertz 1977:165
4. Oakley 1979b:611–12
5. Barbara Rothman 1982:34, 45. See also Osherson and AmaraSingham 1981, Manning and Fabrega 1973:283, 301, Berliner 1982.
6. Wertz and Wertz 1977:40; see also Donnison 1977:42–43
7. On this point, see Rothschild 1981.
8. O'Brien 1981:140ff provides a theoretical and historical discussion of how reproduction has been denied.
9. Noble 1984:x
10. Noble 1984:44
11. Cohen and Estner 1983:173
12. Pritchard et al. 1985:311
13. Braverman 1974
14. O'Driscoll and Foley 1983:5
15. Niswander 1981:207
16. Quoted in Pritchard et al. 1985:314
17. O'Driscoll and Foley 1983:5
18. Rothman 1982:269
19. Ellis and Beckmann 1983:224–29

20. Friedman 1982:166; Ellis and Beckmann 1983:502–3; Iffy and Kaminetzky 1981:819
21. Ellis and Beckmann 1983:501
22. Ellis and Beckmann 1983:501–6
23. Friedman 1982:173
24. Niswander 1981:204, 209
25. Eastman 1950:326
26. Pritchard et al. 1985:307
27. Eastman 1950:326
28. Pritchard and MacDonald 1980, Pritchard et al. 1985
29. Niswander 1981:184
30. Haire 1972:13; Newton 1968:1096–1102
31. Pritchard et al. 1985:643
32. Pritchard et al. 1985:643, emphasis in original.
33. There are complex issues involved in deciding whether the dichotomy (or range) "voluntary-involuntary" adequately captures the variety and sense of human actions. I do not take up these issues here, but see Anscombe 1963:89–90 for a relevant discussion.
34. Benson 1982:905–7; Niswander 1981:207; Pritchard et al. 1985:644
35. Blauner 1964:147–56
36. Pritchard et al. 1985:311
37. See also Pritchard et al. 1985:338.
38. Oakley 1979a:102
39. Drife 1983:239
40. Gellman et al. 1983:2935; Ellis and Beckmann 1983:507
41. Oakley 1984:204–5. Quotations are from O'Driscoll and Meagher 1980:140, 142
42. Marieskind 1983:189
43. Weekes 1983:476
44. Jones 1976:527
45. Blakeslee 1985
46. Hibbard 1976:804, discussion quoting Kroener
47. Barker-Benfield 1976:288
48. Wertz and Wertz quoting DeLee 1977:142–43
49. Rothman 1982:48
50. Meillassoux 1981; Hindess and Hirst 1975. Recent efforts to "understand the inter-relationships between production and reproduction as part of a single process" as they are transformed through history (Beechey 1979:79) include Mackintosh 1977, Edholm et al. 1977, Petchesky 1983, Bridenthal 1976, and Garnsey 1978.
51. Williams 1979:147
52. Oxford English Dictionary 1933:5–6
53. Marx 1967a:102
54. Winner 1977:230
55. Mitchell 1971:108

Chapter 5: Self and Body Image

1. Emerson 1970
2. "Birth report" refers to written accounts of women's birth experiences, often requested by childbirth educators.
3. Schilder 1935:139. I am grateful to Sidney Mintz for telling me about Schilder's work.
4. Schilder 1935:159
5. Schilder 1935:139
6. Schilder 1935:166
7. Schilder 1935:166
8. Scott 1949:139
9. Lakoff and Johnson 1980:4
10. Lakoff and Johnson 1980:7–8
11. Weekes 1983:472
12. Niswander 1981:215
13. Ott et al. 1977
14. Minkoff and Schwarz 1980:36
15. O'Driscoll and Foley 1983
16. Snijders 1973
17. Amiel 1982
18. Franks and Johnstone 1982; Russell 1982
19. Harris 1985:3247; Weekes 1983:476
20. Young 1982:102
21. Marut and Mercer 1979; Cohen 1977; Doering and Entwistle 1981; Cranley et al. 1983. See Chapter 8 for more discussion of cesarean sections.
22. Royall 1983:128
23. Affonso and Stichler 1978:88
24. Wilson and Hovey 1980:24
25. Marut 1978:206
26. Affonso and Stichler 1978:90
27. Affonso and Stichler 1981:55
28. Hardwick 1983:35–36
29. Affonso and Stichler 1978:90
30. Meyer 1981:112
31. Wilson and Hovey 1980:180
32. Trowell 1982
33. Donovan 1977:194
34. Wilson and Hovey 1980:144
35. Marut 1978:203
36. Royall 1983:104
37. Marut and Mercer 1979:261; Cohen and Estner 1983:38
38. Klaus and Kennell 1976
39. Norwood 1984:175
40. On bonding, see Arney 1982:172–73.

41. Karmel 1959:42
42. Walton 1976:143
43. Matria and Mullen 1978:27, 28
44. Todd 1981:3
45. Walton 1976:132
46. Lamaze 1965:16, 13
47. Lamaze 1965:137. Suzanne Arms has pointed out that the Lamaze method gives a woman control over her body but that she remains separate from the "sensations, smells and sights of her body giving birth" (1975:146).
48. Wood 1973
49. Smith-Rosenberg 1974
50. Ehrenreich and English 1978:127
51. Rich 1976:171–72

Chapter 6: Menstruation, Work, and Class

1. Whistnant et al. 1975 show how commercial materials portray menstruation as a hygienic crisis.
2. This disgust over bodily effluvia and concern with keeping clean, pervasive as they are today, may not have always been so. See Elias 1978 for a historian's view of how and when concern with cleanliness of bodily functions like spitting and nose-blowing came into being.
3. See Koff et al. 1981 for more on the intense apprehension women have that they will be discovered to be menstruating.
4. Shapiro-Perl 1984:201
5. Chavkin 1984:196
6. Shapiro-Perl 1984:201, 212. The subject of how workers' access to break time or to the bathroom at any time is determined is complex. Amount of time off varies by trade (Blauner 1964:26, 43, 69, 70) and has been a focus of union struggles (Montgomery 1979). Bell 1956:5 discusses how the strictly controlled rhythm of time during the paid workday controls the workers' time off as well.
7. Butler 1909:312–13
8. Hourwich and Palmer 1936:30
9. Kessler-Harris 1982:241. See also Levine 1924:20, 172, 467 on sanitary conditions in New York shops.
10. Richardson 1905:95
11. Morrison 1973:23–24
12. Van Vorst and Van Vorst 1903:131; Hourwich and Palmer 1939:113; Parker 1922:51; Seifer 1976:50–51
13. Penny 1870:302
14. Campbell 1887:181
15. Campbell 1887:181

16. Parker 1922:67, 134, 37
17. Hourwich and Palmer 1936:84
18. Kessler-Harris 1982:291
19. Porter 1969:22
20. Shorter 1982:287–88
21. Skultans 1970:648
22. de Beauvoir 1953:138; Shorter 1982:288. Some have argued that taboos originate in an understanding that women are an actual danger to society once each month (Darga et al. 1981)!
23. It does not follow that *men* would do this work. It is more probable that nonmenstruating women would take over for a time. However, Okely 1975 says that Gypsy men go to fish-and-chips shops when their wives are menstruating.
24. See Topley 1975:71 for restrictions on Chinese silk industry workers, Shorter 1982:2887 on the wine industry in France, and Porter 1969:22 on dairying in England.
25. Ong 1983 discusses how Malay women villagers, entering factories for the first time, have to learn to use (and for the first time have the money to buy) sanitary napkins. They "need the protection because of the long hours confined at work" (Ong 1983:55). It would be interesting to reconstruct the history of the manufacture and marketing of sanitary napkins and tampons in the west in relation to factory work for women.
26. Luker 1984:158–91
27. Adams and Winston 1980:27, 28
28. Adams and Winston 1980:33
29. Sullerot 1971:213. See also Chavkin 1984.
30. Luker 1984:200
31. Especially among sisters, the date of first menstruation can become a bone of contention. A woman whose sister had a kidney infection when she was nine and started to pass blood, was told by her mother that her sister's period might have started. "I fell apart, completely fell apart. I was stricken at the thought that she would become a woman before I did" (Meg O'Hara).
32. See Ernster 1975 on women's secret language to describe menstruation.
33. It is not that early pregnancy is desired per se (Ladner 1971:226) but that the thought of it does not represent as intense a threat to future career plans as for middle-class women.
34. Stack 1977
35. Ladner 1971:212
36. Rubin 1976:49–68
37. Among working-class girls in the late nineteenth and early twentieth centuries, sex before marriage was regarded as the primary resource girls had to trade for men's monetary and social favors (Peiss 1983:84).
38. Luker 1984:200, 202, 204, 213

39. Ladner 1971:262–65; 1984

40. Whistnant and Zergans 1975 discuss how middle-class women are unable to assimilate biological accounts.

41. Some have urged us to revel in menstrual blood and make it a matter of spiritual delight by developing new rituals, "bleed-ins" (Delaney et al. 1976).

42. Of course, not all women menstruate every month. Low body fat from exercise or diet as well as many other factors can cause menstruation to stop, for a short or long time. What we share as culturally defined women is being the ones who menstruate, other things being equal.

Chapter 7: Premenstrual Syndrome, Work Discipline, and Anger

1. I am indebted to Thomas Buckley and Alma Gottlieb, who solicited the original version of this essay for their volume *Blood Magic*.

2. Lever 1981:108. Other examples of uniformly negative symptomatology are Halbreich and Endicott 1982 and Dalton 1983.

3. Lever 1981:1

4. Ad for a drug in *Dance* magazine, Jan. 1984:55

5. Few accounts reject the description of PMS as a disease. One that does is Witt, who prefers the more neutral term "condition" (1984:11–12). It is also relevant to note that I do not experience severe manifestations of PMS, and there is a possibility for that reason that I do not give sufficient credit to the medical model of PMS. I have, however, experienced many similar manifestations during the first three months of each of my pregnancies, so I have some sense that I know what women with PMS are talking about.

6. Lever 1981:47

7. Lever 1981:2, 1. Other estimates are "up to 75%" (Southam and Gonzaga 1965:154) and 40% (Robinson, Huntington and Wallace 1977:784).

8. Smith-Rosenberg 1974:27

9. Meigs 1854:381

10. Clarke 1873:130

11. Ehrenreich and English 1973:48

12. Clarke 1873:133–34

13. Hollingworth 1914:93

14. Jacobi 1877:232

15. Frank 1931

16. Frank 1931:1054

17. Benedek and Rubenstein 1939 II:461

18. Ivey and Bardwick 1968:344

19. Paige 1971:533–34

20. Frank 1931:1053

21. Kessler-Harris 1982:219, 259, 254–55

22. Seward 1934; McCance 1937; Billings 1934; Brush 1938

23. Altmann 1941; Anderson 1941; Novak 1941; Brinton 1943; Percival 1943

24. Seward 1944:95

25. Kessler-Harris 1982:295

26. Dalton and Greene 1953

27. Weiner 1985:118

28. Kessler-Harris 1982:311–16; Weiner 1985:89–97, 112–18

29. Kessler-Harris 1982:316–18

30. Laws 1983:25. I am working on a more extensive study of the literature on menstruation and industrial work from late nineteenth century to the present, including publications in industrial hygiene as well as medicine and public health.

31. Birke and Gardner 1982:ii

32. Robinson et al. 1977:784

33. Birke and Gardner 1982:24

34. Lever 1981:91; Dalton 1983:102

35. Marx 1967a:412

36. Braverman 1974, Melman 1983. See Edwards 1979:98–104 for factors that limited the impact of scientific management in industry.

37. Braverman 1974: Part 4

38. Sokolov 1984

39. Foucault 1979:139

40. Mullings 1984:131 points to the higher frequency of these hazards among minority and working-class women.

41. Debrovner 1982:11

42. Lever 1982:20

43. Witt 1984:129

44. Vanek 1974; Scott 1980; Cowan 1983:208

45. Braverman 1974:173–75

46. Gilbreth and Carey 1948:2

47. de Beauvoir 1952:425

48. Oakley 1974:45

49. Ballantyne 1975:114

50. Dalton 1983:80

51. Dalton 1983:82

52. Gottlieb 1982:44

53. Dalton 1983:100. These figures are still being cited in major newspapers (Watkins 1986).

54. Parlee 1973:461–62

55. Golub 1976; Sommer 1973; Witt 1984:160–62

56. Both quoted in Weideger 1977:48

57. Birke and Gardner 1982:23

58. Harrison 1984:16–17

59. Witt 1984:150
60. Witt 1984:151
61. Birke and Gardner 1982:25; Bernstein 1977
62. Powers 1980 suggests that the association generally made between menstruation and negative conditions such as defilement may be a result of *a priori* western notions held by the investigator. She argues that the Oglala, Plains Indians, have no such association. This does not mean that menstruation is *never* regarded negatively, of course (see Price 1984:21–22).
63. Buckley 1982:49
64. Harrison 1984:44
65. *Baltimore City Paper* 20 April 1984:39
66. Birke and Gardner 1982:25
67. Lever 1981:61
68. Lever 1981:63
69. Lauersen and Stukane 1983:18
70. Lever 1981:25
71. Lever 1981:28
72. Lever 1981:68
73. Angier 1983:119
74. Lauersen and Stukane 1983:80
75. Letter in *PMS Connection* 1982, 1:3. Reprinted by permission of PMS Action, Inc., Irvine, CA.
76. Lewis 1971:116
77. Weideger 1977:10
78. Harrison 1984:50
79. Frye 1983:91
80. Lever 1981:61
81. Harrison 1984:36
82. On isolation of housewives, see Gilman 1903:92.
83. Rich 1976:285
84. Shuttle and Redgrove 1978:58, 59
85. Griffin 1978:185
86. Gramsci 1971:333
87. Fanon 1963; Genovese 1976:647
88. Lorde 1981:9, 8
89. Dalton 1983:80; Harrison 1984:17; Halbreich and Endicott 1982:251, 255, 256
90. Dalton 1983:80
91. *PMS Connection* 1984, 3:4
92. Fox-Genovese 1984:272–73
93. Sanders 1981
94. Witt 1984:205, 208; Herrmann and Beach 1978
95. Thompson 1967
96. Rossi and Rossi 1974:32
97. Braverman 1974:32

98. Steinem 1981:338

Chapter 8: Birth, Resistance, Race, and Class

The epigraph to this chapter is reprinted by permission of Marion Cohen and of *Mothering* magazine, vol. 16, published by Mothering Publications; all rights reserved.

1. Noble 1984:33
2. Downer 1984:425; Cassidy-Brinn et al. 1984
3. See Cassidy-Brinn et al. 1984 for a more complete list.
4. Rothman 1982:269,270
5. Cohen and Estner 1983:183
6. Noble 1984:345
7. Noble 1984:349
8. Quoted in Wertz and Wertz 1977:39
9. Ranney 1973
10. Noble 1984:62, 65
11. Hubbard 1984:337; Murphy 1984; Corea 1985
12. Rothman 1984:27
13. Gallagher 1984; Hubbard 1982
14. Hauseknecht and Heilman 1982:196. On telemetered fetal monitoring see Arney 1982:149.
15. Etzioni 1979:59
16. Grossman 1971:48
17. Noble 1984:48. Oakley 1984: 283 also makes this point.
18. Rosengren and Devault 1963; Starr 1984:446–47; Kinkead 1980:76; Ehrenreich and Ehrenreich 1971
19. This process is complicated because some doctors are also owner-managers of health corporations (Starr 1984:443).
20. *The* (Baltimore) *Sun* 5 March 1985
21. Eastman 1950:324
22. Hellman and Pritchard 1971:349
23. Pritchard and MacDonald 1980:369–93
24. Pritchard et al. 1985:299
25. Pritchard et al. 1985:299
26. Lerner and Stutz 1975
27. Conover 1973
28. Shapiro et al. 1968:145, 150–51
29. Shapiro et al. 1968:66
30. Shapiro et al. 1968:67
31. Shapiro et al. 1968:150–51
32. Mullings 1984:121
33. Fuchs 1974:34
34. Mullings 1984:123, 129

35. Eyer and Sterling 1977:16–17
36. Karasek et al. 1981; Nelson 1983
37. U.S. Department of Health and Human Services 1981:164, 165
38. Hurst and Summey 1984; Gleicher 1984:3273
39. U.S. Department of Health and Human Services 1981
40. Hurst and Summey 1984:629
41. Shapiro et al. 1968:189, 103
42. Marieskind 1979:115–17
43. Gibbons 1976:189
44. U.S. Department of Health and Human Services 1981:163; Gibbons 1976:24
45. Wegman 1984:987
46. Gibbons 1976:180, 69
47. Minkoff and Swartz 1980. In a study of the rates of induction of labor in England, Ann Cartwright 1979:112 found that more middle-class women than working-class women felt they could refuse induction.
48. Segal 1985
49. Scully 1980:130, 135

Chapter 9: The Creation of New Birth Imagery

1. Young 1982:449–51
2. Young 1982:209–23
3. Cohen and Estner 1983:120
4. Cohen and Estner 1983:120–21
5. Baldwin 1979:81
6. Baldwin 1979:81, 135
7. Gaskin 1977:239
8. Gaskin 1977:87
9. Panuthos 1984:484
10. Peterson 1984:52
11. Cohen and Estner 1983:223
12. Rothman 1982:20
13. Moran 1983:17
14. Moran 1984:10
15. Moran 1983:2
16. Moran 1984:9–10
17. Koeske 1983:13
18. Koeske 1983:13
19. Odent 1981:7
20. Odent 1981:11
21. Odent 1981:15
22. Odent 1984:12–13

23. Bradley 1977 makes the same assumptions in his advocacy of husband-coached childbirth.

24. See Ortner 1974 on gender in relation to the split between nature and culture.

Chapter 10: Menopause, Power, and Heat

1. McKinlay and Jefferys 1974
2. Lock 1986:35
3. Davis 1986:80
4. Reitz 1977:27
5. Gray 1981
6. Goffman 1967:97
7. Goffman 1967:110
8. Edelmann 1981:127
9. Ryle 1968
10. Weinberg 1968:384
11. Budoff 1983:25–26
12. Reitz 1977:30
13. Kaufert and Syrotuik 1981; Kaufert 1982
14. Levit 1963:31
15. Frey 1981:31, 33. See also the earlier study by Goodman et al. 1977 on how menopausal symptoms are less widespread than previously believed.
16. van Keep and Kellerhals 1974 found this correlation in a Swiss study; but McKinlay and Jefferys 1974 could not find any particular relationship between menopausal complaints and social class.
17. Flint 1979:43; Neugarten and Kraines 1965
18. Severne 1979:105; Lomax 1982
19. Special Task Force to the Secretary of Health, Education, and Welfare 1973
20. Special Task Force to the Secretary of Health, Education, and Welfare 1973:77
21. Lakoff and Johnson 1980:15
22. Lakoff and Johnson 1980:16
23. Lakoff and Johnson 1980:17
24. See Lakoff and Kövecses 1987 for an extended discussion of the American cognitive model of anger.
25. Reitz 1977:79
26. Reitz 1977:31
27. Neugarten et al. 1968; Muhlenkamp et al. 1983; Crawford and Hooper 1973
28. I do not deal with the controversy over whether to prescribe estrogen to women during menopause because the outlines of the debate are clear from the literature (McCrea and Markle 1984) and because hardly any of the

women we interviewed were taking estrogen or indicated it had been or was an issue for them.

29. Neugarten et al. 1968 also found a discrepancy between younger and older women's attitudes toward menopause, the younger women being far more negative.

30. O'Neill 1982:29. This kind of attitude toward menopausal women is still found in recent publications, despite a vigorous feminist critique. For critiques, see MacPherson 1981, Posner 1979, and Bruck 1979.

31. O'Neill 1982:7–8

32. O'Neill 1982:29

33. Rauramo, quoted in Lauritzen and van Keep 1977

34. See McCrea 1983.

35. Kathy Kozachenko, quoted in Bart and Grossman 1978:351–52

Chapter 11: Class and Resistance

1. Mitchell 1971:22

2. Mullings 1984:134. See also Bryce-Laporte and Thomas 1976:xxxv.

3. See Dwyer 1978, who documents this for a Middle Eastern society.

4. Abercrombie et al. 1980:2. See Mann's summary of and references to the social science literature claiming that unskilled workers lack the conception of an alternative society and lack the confidence in their own power necessary for revolution (1973:24–33).

5. Rowbotham 1973:30

6. Firestone 1973:149. Quoted in Sutton et al. 1976:186. Arguing from a psychoanalytic perspective, Dinnerstein draws the very different conclusion that women are more likely to challenge the status quo than men because of the onus that falls on women of being responsible for everything, which women find unspeakable and unjust 1976:234.

7. Abercrombie et al. 1980:140–44

8. Abercrombie et al. 1980:70

9. Therborn 1980:17, 94–97, 95

10. U.S. Department of Health and Human Services 1981

11. Jaggar 1983:370–71

Chapter 12: The Embodiment of Oppositions

1. Barnett and Silverman 1979:35; Lasch 1984:195–96

2. Marcuse 1964:23

3. Sennett and Cobb 1973

4. Wittgenstein 1969

5. Wittgenstein 1969:161

6. Bernstein 1971:61

7. Hymes 1980:40
8. Hymes 1980:41. See Brittan and Maynard 1984:88–89 for the argument that middle-class socialization practices and exposure to education allow for a greater degree of "reflexive questioning."
9. Hymes 1980:127–28, 131, 134
10. Lewontin et al. 1984:133–34
11. Whorf 1956:62
12. Thompson 1967
13. Taub and Schneider 1982
14. Mullings 1984
15. Chodorow 1978
16. Hartsock 1983:297
17. Hartsock 1983:298
18. Hartsock 1983:299
19. Smith 1979:166–68
20. Jaggar 1983:371
21. Brittan and Maynard 1984:87
22. Aptheker 1982:113–28; Blauner 1972:23–24; Davis 1981:237–38
23. Rich 1978
24. Rubin 1976

References

Abercrombie, Nicholas, Stephen Hill, and Bryan S. Turner 1980 *The Dominant Ideology Thesis*. London: George Allen and Unwin Ltd.

Adams, Carolyn Teich and Kathryn Teich Winston 1980 *Mothers at Work: Public Policies in the United States, Sweden, and China*. New York: Longman.

Affonso, Dyanne D. and Jaynelle F. Stichler 1978 "Exploratory Study of Women's Reactions to Having a Cesarean Birth." *Birth and the Family Journal* 5(2):88–94.

Altmann, M. 1941 "A Psychosomatic Study of the Sex Cycle in Women." *Psychosomatic Medicine* 3:199–225.

Amiel, Gerard J. 1982 "Breech: Vaginal Delivery or Caesarean Section?" *British Medical Journal* 285:1275.

Anderson, M. 1941 "Some Health Aspects of Putting Women to Work in War Industries." *Industrial Hygiene Foundation 7th Annual Meeting:* 165–69.

Angier, Natalie with Janet Witzleben 1983 "Dr. Jekyll and Ms. Hyde." *Reader's Digest* 43:119–21.

Anscombe, G. E. M. 1963 *Intention*. Ithaca, NY: Cornell University Press.

Aptheker, Bettina 1982 *Woman's Legacy: Essays on Race, Sex, and Class in American History*. Amherst, MA: The University of Massachusetts Press.

Ardener, Shirley 1978 "Introduction: The Nature of Women in Society," pp. 9–48 in *Defining Females: The Nature of Women in Society*. New York: John Wiley.

Arms, Suzanne 1975 *Immaculate Deception: A New Look at Women and Childbirth in America*. Boston: Houghton Mifflin.

Arney, William Ray 1982 *Power and the Profession of Obstetrics*. Chicago: University of Chicago Press.

Aronowitz, Stanley 1973 *False Promises: The Shaping of American Working Class Consciousness*. New York: McGraw-Hill.

Baldwin, Rahima 1979 *Special Delivery: The Complete Guide to Informed Birth*. Millbrae, CA: Les Femmes.

Ballantyne, Sheila 1975 *Norma Jean the Termite Queen*. New York: Penguin.

Barker-Benfield, G. J. 1976 *The Horrors of the Half-Known Life: Male Attitudes Toward Women and Sexuality in Nineteenth-Century America*. New York: Harper and Row.

Barnett, Steve and Martin G. Silverman 1979 *Ideology and Everyday Life: Anthropology, Neomarxist Thought, and the Problem of Ideology and the Social Whole*. Ann Arbor, MI: The University of Michigan Press.

Bart, Pauline 1976 "Alienation, Women and Health," pp. 199–215 in *Alienation in Contemporary Society,* Roy S. Bryce-Laporte and Claudewell S. Thomas, eds. New York: Praeger.

Bart, Pauline B. and Marilyn Grossman 1978 "Menopause," pp. 337–54 in *The Woman Patient: Medical and Psychological Interfaces,* Malkah T. Notman and Carol C. Nadelson, eds. New York: Plenum.

Bartky, Sandra Lee 1982 "Narcissism, Femininity and Alienation." *Social Theory and Practice* 8(2):127–43.

Beckmann, Charles R. B. and Ellis, Jeffrey W. 1983 *A Clinical Manual of Obstetrics*. Norwalk, CT: Appleton-Century-Crofts.

Beechey, Veronica 1979 "On Patriarchy." *Feminist Review* 3:66–82.

Bell, Daniel 1956 *Work and Its Discontents*. Boston: Beacon Press.

Benedek, Therese and Boris B. Rubenstein 1939 "The Correlations Between Ovarian Activity and Psychodynamic Processes. I. The Ovulative Phase; II. The Menstrual Phase." *Psychosomatic Medicine* 1(2):245–70; 1(4):461–85.

Benenson, Harold Berger 1980 "The Theory of Class and Structural Developments in American Society: A Study of Occupational and Family Change, 1945–70." Unpublished dissertation, New York University.

Benson, Ralph C. 1982 *Current Obstetric and Gynecologic Diagnosis and Treatment*. Los Altos, CA: Lange Medical Publishers.

Berliner, Howard 1975 "A Larger Perspective on the Flexner Reforms." *International Journal of Health Services* 5(4):573–92.

———. 1982 "Medical Modes of Production," pp. 162–71 in *The Problem of Medical Knowledge: Examining the Social Construction of Medicine,* Peter Wright and Andrew Treacher, eds. Edinburgh: Edinburgh University Press.

Bernstein, Barbara Elaine 1977 "Effect of Menstruation on Academic Women." *Archives of Sexual Behavior* 6(4):289–96.

Bernstein, Basil 1971 *Class, Codes and Control: Volume 1, Theoretical Studies Towards a Sociology of Language*. London: Routledge and Kegan Paul, Ltd.

Berreman, Gerald D. 1972 "Race, Caste, and Other Invidious Distinctions in Social Stratification." Reprinted, pp. 500–36 in *Anthropology for the Eighties,* Johnetta B. Cole, ed. [1982] New York: Macmillan Publishing Co.

Billings, Edward G. 1933 "The Occurrence of Cyclic Variations in Motor

Activity in Relation to the Menstrual Cycle in the Human Female." *Bulletin of Johns Hopkins Hospital* 54:440–54.
Birke, Lynda and Katy Gardner 1982 *Why Suffer? Periods and Their Problems.* London: Virago.
Blakeslee, Sandra 1985 "Doctors Debate Surgery's Place in the Maternity Ward." *New York Times,* March 24.
Blauner, Robert 1964 *Alienation and Freedom: The Factory Worker and His Industry.* Chicago: University of Chicago Press.
———. 1972 *Racial Oppression in America.* New York: Harper and Row.
Boston Women's Health Book Collective 1984 *The New Our Bodies, Our Selves.* New York: Simon and Schuster.
Bradley, Robert A. 1977 *Husband-Coached Childbirth.* New York: Harper and Row.
Braverman, Harry 1974 *Labor and Monopoly Capital.* New York: Monthly Review Press.
Bridenthal, Renate 1976 "The Dialectics of Production and Reproduction in History." *Radical America,* 10 March 1976:3–11.
Brinton, Hugh P. 1943 "Women in Industry," pp. 395–419 in *Manual of Industrial Hygiene and Medical Service in War Industries,* National Institutes of Health, Division of Industrial Hygiene. Philadelphia: W. B. Saunders.
Brittan, Arthur and Mary Maynard 1984 *Sexism, Racism and Oppression.* Oxford: Basil Blackwell.
Brown, E. Richard 1979 *Rockefeller Medicine Men: Medicine and Capitalism in America.* Berkeley: University of California Press.
Bruck, Connie 1979 "Menopause." *Human Behavior* 8(4):38–46.
Brush, A. L. 1938 "Attitudes, Emotional and Physical Symptoms Commonly Associated with Menstruation in 100 Women." *American Journal of Orthopsychiatry* 8:286–301.
Bryce-Laporte, Roy S. and Claudewell S. Thomas 1976 Introduction in *Alienation in Contemporary Society: A Multidisciplinary Examination.* New York: Praeger Publishers.
Buckley, Thomas 1982 "Menstruation and the Power of Yurok Women: Methods in Cultural Reconstruction." *American Ethnologist* 9(1):47–60.
Buckley, Thomas, and Alma Gottlieb, eds. 1988 *Blood Magic: Explorations in the Anthropology of Menstruation.* Berkeley: University of California Press.
Budoff, Penny Wise 1984 *No More Hot Flashes and Other Good News.* New York: Warner Books.
Bullough, Vern L. 1975 "Sex and the Medical Model." *The Journal of Sex Research* 11(4):291–303.
Burawoy, Michael 1979 *Manufacturing Consent.* Chicago: University of Chicago Press.
Burrow, James G. 1977 *Organized Medicine in the Progressive Era: The Move Toward Monopoly.* Baltimore: Johns Hopkins University Press.
Butler, Elizabeth Beardsley 1909 *Women and the Trades: Pittsburgh, 1907–08.*

Vol. 1 of *The Pittsburgh Survey,* Paul Underwood Kellogg, ed. New York: Russell Sage Foundation.

Campbell, Helen 1887 *Prisoners of Poverty: Women Wage Workers.* Boston: Roberts Bros.

Cannon, Walter B. 1953 *Bodily Changes in Pain, Hunger, Fear and Rage: An Account of Recent Researches into the Function of Emotional Excitement.* Boston: Charles T. Branford Company.

Cartwright, Ann 1979 *The Dignity of Labour? A Study of Childbearing and Induction.* London: Tavistock.

Cassidy-Brinn, Francie Hornstein, and Carol Downer 1984 *Woman-Centered Pregnancy and Birth.* Pittsburgh: Cleis Press.

Chavkin, Wendy 1984 "Walking a Tightrope: Pregnancy, Parenting, and Work," pp. 196–213 in *Double Exposure: Women's Health Hazards on the Job and at Home,* Wendy Chavkin, ed. New York: Monthly Review Press.

Chodorow, Nancy 1978 *The Reproduction of Mothering: Psychoanalysis and the Sociology of Gender.* Berkeley: University of California Press.

Clarke, Anne E. and Diane N. Ruble 1978 "Young Adolescents' Beliefs Concerning Menstruation." *Child Development* 49:231–234.

Clarke, Edward H. 1873 *Sex in Education; or a Fair Chance for the Girls.* Boston: James R. Osgood and Co.

Cohen, Nancy Wainer 1977 "Minimizing Emotional Sequellae of Cesarean Childbirth." *Birth and the Family Journal* 4(3):114–19.

Cohen, Nancy Wainer and Lois J. Estner 1983 *Silent Knife: Cesarean Prevention and Vaginal Birth After Cesarean (VBAC).* South Hadley, MA: Bergin and Garvey.

Conover, P. W. 1976 "Social Class and Chronic Illness." *International Journal of Health Services* 3:357–68.

Corea, Gena 1985 *The Mother Machine: Reproductive Technologies from Artificial Insemination to Artificial Wombs.* New York: Harper and Row.

Cowan, Ruth Schwartz 1983 *More Work for Mother: The Ironies of Household Technologies from the Open Hearth to the Microwave.* New York: Basic Books.

Cranley, Mecca S., Kathleen J. Hedahl, and Susan H. Pegg 1981 "Women's Perceptions of Vaginal and Cesarean Deliveries." *Nursing Research* 32(1):10–15.

Crawford, Marion and Douglas Hooper 1973 "Menopause, Aging and the Family." *Social Science and Medicine* 7:469–82.

Crawford, Patricia 1981 "Attitudes to Menstruation in Seventeenth-Century England." *Past and Present* 91:47–73.

Currier, Andrew F. 1897 *The Menopause.* New York: Appleton.

Dalton, Katharina 1983 *Once a Month.* Claremont, CA: Hunter House.

Dalton, Katharina and Raymond Greene 1983 "The Premenstrual Syndrome." *British Medical Journal,* May:1016–17

Darga, Linda L., Suzanne Klauenberg, and Karen F. Davis 1981 "The Cultural Model of the Menstrual Cycle." *Michigan Academician* 13(4):475–83.

Davis, Angela Y. 1981 *Women, Race and Class*. New York: Random House.

Davis, Dona Lee 1986 "The Meaning of Menopause in a Newfoundland Fishing Village." *Culture, Medicine and Psychiatry* 10(1):73–94.

de Beauvoir, Simone 1952 *The Second Sex*. New York: Knopf.

Debrovner, Charles, ed. 1982 *Premenstrual Tension: A Multidisciplinary Approach*. New York: Grove Press.

Delaney, Janice, Mary Jane Lupton, and Emily Toth, eds. 1976 *The Curse: A Cultural History of Menstruation*. New York: E. P. Dutton.

Dinnerstein, Dorothy. 1976 *The Mermaid and the Minotaur: Sexual Arrangements and Human Malaise*. New York: Harper and Row.

Donnison, Jean 1977 *Midwives and Medical Men: A History of Inter-Professional Rivalries and Women's Rights*. New York: Schocken.

Donovan, Bonnie 1977 *The Cesarean Birth Experience: A Practical, Comprehensive, and Reassuring Guide for Parents and Professionals*. Boston: Beacon Press.

Dougherty, Molly Crocker 1978 *Becoming a Woman in Rural Black Culture*. New York: Holt, Rinehart and Winston.

Downer, Carol 1984 "Through the Speculum," pp. 419–26 in *Test-tube Women*, Rita Arditti, Renate Duelli Klein, and Shelley Minden, eds. London: Routledge and Kegan Paul, Ltd.

Drife, James Owen 1983 "Kielland or Caesar?" *British Medical Journal* 287:309–10.

Dwyer, Daisy Hilse 1978 "Ideologies of Sexual Inequality and Strategies for Change in Male-Female Relations." *American Ethnologist* 5:227–40.

Eastman, Nicholson J. 1950 *Williams Obstetrics*, 10th Edition. New York: Appleton-Century-Crofts.

Edelmann, Robert J. 1981 "Embarrassment: The State of Research." *Current Psychological Reviews* 1:125–38.

Edholm, Felicity, Olivia Harris, and Kate Young 1977 "Conceptualizing Women." *Critique of Anthropology* 3:101–30.

Edwards, Richard 1979 *Contested Terrain: The Transformation of the Workplace in the Twentieth Century*. New York: Basic Books.

Ehrenreich, Barbara and Deirdre English 1973 *Complaints and Disorders: The Sexual Politics of Sickness*. Old Westbury, New York: The Feminist Press.

———. 1978 *For Her Own Good: 150 Years of the Experts' Advice to Women*. New York: Doubleday.

Ehrenreich, Barbara and John Ehrenreich 1971 *The American Health Empire: Power, Profits, and Politics*. New York: Vintage.

Elias, Norbert 1978 *The Civilizing Process: The History of Manners*. Edmund Jephcott, trans. New York: Urizen Books.

Ellis, Havelock 1904 *Man and Woman*. London: Walter Scott.

Ellis, Jeffrey W. and Charles R. B. Beckmann 1983 *Clinical Manual of Obstetrics*. Appleton-Century-Crofts.

Elshtain, Jean Bethke 1981 *Public Man, Private Woman*. Princeton, NJ: Princeton University Press.

Emerson, Joan P. 1970 "Behavior in Private Places: Sustaining Definitions of Realities in Gynecological Examinations," pp. 74–97 in *Recent Sociology No. 2: Patterns of Communicative Behavior,* Hans Peter Dreitzel, ed. New York: Macmillan Publishing Co.

Engels, Frederick 1972 *The Origin of the Family, Private Property and the State.* Introduction by Eleanor Burke Leacock. New York: International Publishers.

Entwhistle, Doris R., and Susan G. Doering 1981 *The First Birth: A Family Turning Point.* Baltimore: Johns Hopkins University Press.

Ernster, Virginia L. 1975 "American Menstrual Expressions." *Sex Roles* 1(1):3–13.

Etzioni, Amitai 1973 *Genetic Fix.* New York: Macmillan Publishing Co.

Ewbank, Thomas 1855 *The World a Workshop: or the Physical Relationship of Man to the Earth.* New York: Appleton.

Eyer, Joseph and Peter Sterling 1977 "Stress-Related Mortality and Social Organization." *The Review of Radical Political Economics* 9(Spring):1–44.

Fanon, Frantz 1963 *The Wretched of the Earth.* New York: Grove Press.

Federation of Feminist Women's Health Centers 1981 *A New View of a Woman's Body.* New York: Simon and Schuster.

Fee, Elizabeth 1976 "Science and the Woman Problem: Historical Perspectives," pp. 175–223 in *Sex Difference: Social and Biological Perspectives,* Michael S. Teitelbaum, ed. New York: Doubleday.

———. 1981 "Is Feminism a Threat to Scientific Objectivity?" *International Journal of Women's Studies* 4:378–92.

Finifter, Ada W., ed. 1972 *Alienation and the Social System.* New York: John Wiley.

Firestone, Sulamith 1970 *The Dialectic of Sex: The Case for Feminist Revolution.* New York: William Morrow Co.

Fisher, Marvin 1966 "Melville's 'Bell-Tower': A Double Thrust." *American Quarterly* 18:200–207.

———. 1967 *Workshops in the Wilderness: The European Response to American Industrialization, 1830–1860.* New York: Oxford University Press.

Flint, M. P. 1979 "Sociology and Anthropology of the Menopause," pp. 1–8 in *Female and Male Climacteric: Current Opinion 1978.* P. A. van Keep, D. M. Serr, and R. B. Greenblatt, eds. Lancaster, England: MTP Press.

Foucault, Michel 1979 *Discipline and Punish: The Birth of the Prison.* New York: Vintage.

Fox-Genovese, Elizabeth 1982 "Gender, Class and Power: Some Theoretical Considerations." *The History Teacher* 15(2):255–76.

Frank, Jerome D. 1981 "Holistic Medicine: A View from the Fence." *The Johns Hopkins Medical Journal* 149:222–27.

Frank, Robert T. 1931 "The Hormonal Causes of Premenstrual Tension." *Archives of Neurology and Psychiatry* 26:1053–57.

Franks, P., V. Paily, and M. J. Johnstone 1982 "Breech: Vaginal Delivery or Caesarean Section?" *British Medical Journal* 285:1426.

Frey, Karen A. 1981 "Middle-aged Women's Experience and Perceptions of Menopause." *Women and Health* 6(1–2):25–36.

Friedman, Emanuel A. 1982 "Arrest Disorders," pp. 172–73 in *Obstetrical Decision Making: By the Staff of Beth Israel Hospital, Boston,* Emanuel A. Friedman, M.D., ed. Trenton, NJ: Decker.

Friedman, Emanuel A. and Marlene Sachtleben 1965 "Relation of Maternal Age to the Course of Labor." *American Journal of Obstetrics and Gynecology* 91(7):915–24.

Friedson, Eliot 1970 *Profession of Medicine: A Study of the Sociology of Applied Knowledge.* New York: Dodd, Mead.

Frye, Marilyn 1983 *The Politics of Reality: Essays in Feminist Theory.* Trumansburg, NY: The Crossing Press.

Fuchs, Victor A. 1974 *Who Shall Live? Health, Economics and Social Choice.* New York: Basic Books.

Gallagher, Janet 1984 "The Fetus and the Law: Whose Life Is It Anyway?" *Ms.* 13(3):62.

Ganong, William F. 1983 *Review of Medical Physiology.* 11th Edition. Los Altos, CA: Lange.

———. 1985 *Review of Medical Physiology.* 12th Edition. Los Altos, CA: Lange.

Garnsey, Elizabeth 1978 "Women's Work and Theories of Class Stratification." *Journal of Sociology* 12(2):223–43.

Gaskin, Ina May 1977 *Spiritual Midwifery.* Summertown, TN: The Book Publishing Co.

Geddes, Patrick and J. Arthur Thompson 1890 *The Evolution of Sex.* New York: Scribner and Welford.

Geertz, Clifford 1983 *Local Knowledge: Further Essays in Interpretive Anthropology.* New York: Basic Books.

Gellman, Elliott, Martin S. Goldstein, Soloman Kaplan, and William J. Shapiro 1983 "Vaginal Delivery after Cesarean Section." *Journal of the American Medical Association* 249(21):2935–37.

Genovese, Eugene D. 1974 *Roll, Jordan, Roll: The World the Slaves Made.* New York: Vintage.

Gibbons, Lillian K. 1976 *Analysis of the Rise in C-Section in Baltimore.* Unpublished Ph.D. dissertation, Johns Hopkins School of Hygiene and Public Health.

Giddens, Anthony 1975 *The Class Structure of the Advanced Societies.* New York: Harper and Row.

Gilbreth, Frank B. and Ernestine Gilbreth Carey 1948 *Cheaper by the Dozen.* New York: Bantam.

Gilman, Charlotte Perkins 1903 *The Home: Its Work and Influence.* Urbana: University of Illinois Press (1972 reprint).

Gilman, Sander L. 1985 *Difference and Pathology: Stereotypes of Sexuality, Race, and Madness.* Ithaca, NY: Cornell University Press.

Gleicher, Norbert 1984 "Cesarean Section Rates in the United States: The

Short-term Failure of the National Consensus Development Conference in 1980." *Journal of the American Medical Association* 252(23):3273–76.

Goffman, Erving 1967 *Interaction Ritual: Essays on Face-To-Face Behavior.* Garden City, NY: Anchor.

Golub, Sharon 1976 "The Effect of Premenstrual Anxiety and Depression on Cognitive Function." *Journal of Personality and Social Psychology* 34(1):99–104.

Good, John Mason 1843 *The Study of Medicine,* V. 2. New York: Harper and Bros.

Goodman, Madeleine, C. J. Stewart and F. Gilbert 1977 "Patterns of Menopause: A Study of Certain Medical and Physiological Variables among Caucasian and Japanese Women Living in Hawaii." *Journal of Gerontology* 32:297–98.

Gottlieb, Alma 1982 "Sex, Fertility and Menstruation among the Beng of the Ivory Coast: A Symbolic Analysis." *Africa* 52(4):34–47.

Gramsci, Antonio 1971 *Prison Notebooks.* New York: International Publishers.

Gray, Madeline 1981 *The Changing Years: The Menopause Without Fear.* New York: Signet.

Greer, Edward 1982 "Antonio Gramsci and Legal Hegemony," pp. 304–9 in *The Politics of Law: A Progressive Critique,* David Kairys, ed. New York: Pantheon.

Griffin, Susan 1978 *Woman and Nature: The Roaring Inside Her.* New York: Harper and Row.

Grossman, Edward 1971 "The Obsolescent Mother." *The Atlantic Monthly,* 227:39–50.

Guyton, Arthur C. 1981 *Textbook of Medical Physiology.* Philadelphia, W. B. Saunders.

———. 1984 *Physiology of the Human Body.* 6th Edition. Philadelphia: Saunders College Publishing.

———. 1986 *Textbook of Medical Physiology.* 7th Edition. Philadelphia: W. B. Saunders.

Haber, Carole 1983 *Beyond Sixty-Five: The Dilemma of Old Age in America's Past.* Cambridge: Cambridge University Press.

Haire, Doris 1972 *The Cultural Warping of Childbirth.* Hillside, NJ: International Childbirth Association.

Halbreich, Uriel and Jean Endicott 1982 "Classification of Premenstrual Syndromes," pp. 243–65 in *Behavior and the Menstrual Cycle,* Richard C. Friedman, ed. New York: Marcel Dekker.

Haraway, Donna 1978 "Animal Sociology and a Natural Economy of the Body Politic." *Signs* 4:21–36.

———. 1979 "The Biological Enterprise: Sex, Mind, and Profit from Human Engineering to Sociobiology." *Radical History Review* 20:206–37.

Hardwick, Monica 1983 "Caesarean Section under Epidural: A Personal Account." *British Medical Journal* 287:35–36.

Harris, Bruce A. 1985 Letter to the Editor. *Journal of the American Medical Association* 253(22):3247.

Harrison, Michelle 1984 *Self-Help for Premenstrual Syndrome*. Cambridge, MA: Matrix Press.

Hartsock, Nancy C. M. 1983 "The Feminist Standpoint: Developing the Ground for a Specifically Feminist Historical Materialism," pp. 283–310 in *Discovering Reality,* Sandra Harding and Merrill B. Hintikka, eds. Dordrecht, Netherlands: D. Reidel.

Hausknecht, Richard and Joan Rattner Heilman 1982 *Having a Cesarean Baby.* New York: Dutton.

Hellman, Louis M. and Jack A. Pritchard 1971 *Williams Obstetrics*. New York: Appleton-Century-Crofts.

Henderson, Randi 1985 "'Just So Fortunate.'" *The* [Baltimore] *Sun,* 13 June.

Herrmann, W. M. and R. C. Beach 1978 "Experimental and Clinical Data Indicating the Psychotropic Properties of Progestogens," *Postgraduate Medical Journal* 54:82–87.

Hibbard, Lester T. 1976 "Changing Trends in Cesarean Section." *American Journal of Obstetrics and Gynecology* 125(6):798–804.

Hindess, Barry and Paul Q. Hirst 1975 *Pre-Capitalist Modes of Production*. London: Routledge and Kegan Paul, Ltd.

Hollingworth, Leta 1914 *Functional Periodicity: An Experimental Study of the Mental and Motor Abilities of Women During Menstruation*. New York: Teacher's College.

Horrobin, David F. 1973 *Introduction to Human Physiology.* Philadelphia: F. A. Davis.

Hourwich, Andria Taylor and Gladys L. Palmer, eds. 1936 *I Am a Woman Worker: A Scrapbook of Autobiographies*. New York: The Affiliated Schools for Workers. (Reprint edition 1974 by Arno Press.)

House of Representatives 1983 "National Organ Transplant Act." *Congressional Record* 129(132). Washington, DC.

Hubbard, Ruth 1982 "Some Legal and Policy Implications of Recent Advances in Prenatal Diagnosis and Fetal Therapy." *Women's Rights Law Reporter* 7(3):201–18.

———. 1984 "Personal Courage Is Not Enough: Some Hazards of Childbearing in the 1980s," pp. 331–55 in *Test-tube Women,* Rita Arditti, Renate Duelli Klein, and Shelley Minden, eds. London: Routledge and Kegan Paul, Ltd.

Hurst, Marsha and Pamela S. Summey 1984 "Childbirth and Social Class: The Case of Cesarean Delivery." *Social Science and Medicine* 18(8):621–31.

Hymes, Dell 1980 *Language in Education: Ethnolinguistic Essays* (Language and Ethnography Series). Washington, DC: Center for Applied Linguistics.

Iffy, Leslie and Harold A. Kaminetzky 1981 *Principles and Practice of Obstetrics and Perinatology.* New York: John Wiley.

Israel, Joachim 1971 *Alienation from Marx to Modern Sociology: A Macrosociological Analysis*. Boston: Allyn and Bacon.

Ivey, Melville E. and Judith M. Bardwick 1968 "Patterns of Affective Fluctuation in the Menstrual Cycle." *Psychosomatic Medicine* 30(3):336–45.

Jacobi, Mary Putnam 1877 *The Question of Rest for Women During Menstruation.* New York: Putnam's Sons.

Jaggar, Alison M. 1983 *Feminist Politics and Human Nature.* Sussex, England: The Harvester Press, Ltd.

Jones, O. Hunter 1976 "Cesarean Section in Present-day Obstetrics." *American Journal of Obstetrics and Gynecology* 126(5):521–30.

Jones, Howard W. and Georgeanna Seegar Jones 1981 *Novak's Textbook of Gynecology.* 10th Edition. Baltimore, MD: Williams and Wilkins.

Karasek, Robert, Dean Baker, Frank Marxer, Anders Ahlbom, and Tores Theorell 1981 "Job Decision Latitude, Job Demands, and Cardiovascular Disease: A Prospective Study of Swedish Men." *American Journal of Public Health* 71(7):694–705.

Karmel, Marjorie 1959 *Thank You, Dr. Lamaze: A Mother's Experience in Painless Childbirth.* Philadelphia: Lippincott.

Kasson, John F. 1976 *Civilizing the Machine: Technology and Republican Values in America 1776–1900.* New York: Penguin.

Kaufert, Patricia A. 1982 "Myth and the Menopause." *Sociology of Health and Illness* 4(2):141–65.

Kaufert, Patricia A. and Penny Gilbert 1986 "Women, Menopause and Medicalization." *Culture, Medicine and Psychiatry* 10(1):7–21.

Kaufert, Patricia and John Syrotuik 1981 "Symptom Reporting at the Menopause." *Social Science and Medicine* 15E:173–84.

Keller, Evelyn Fox 1985 *Reflections on Gender and Science.* New Haven, CT: Yale University Press.

Kessler-Harris, Alice 1982 *Out to Work: A History of Wage-Earning Women in the United States.* New York: Oxford University Press.

Kinkead, Gwen 1980 "Humana's Hard-Sell Hospitals." *Fortune,* 17 Nov.: 68–81.

Klaus, Marshall H. and John H. Kennell, eds. 1976 *Maternal-Infant Bonding.* St. Louis: C. V. Mosby.

Koeske, Randi Daimon 1983 "Lifting the Curse of Menstruation: Toward a Feminist Perspective on the Menstrual Cycle." *Women and Health* 8(2–3): 1–16.

Koff, Elissa, Jill Rierdan and Stacy Jacobson 1981 "The Personal and Interpersonal Significance of Menarche." *Journal of the American Academy of Child Psychiatry* 20:148–58.

Ladner, Joyce A. 1971 *Tomorrow's Tomorrow: The Black Woman.* New York: Doubleday.

Lakoff, George and Mark Johnson 1980 *Metaphors We Live By.* Chicago: University of Chicago Press.

Lakoff, George and Zoltán Kövecses 1987 "The Cognitive Model of Anger Inherent in American English," pp. 195–221 in *Cultural Models in Language and Thought,* Dorothy Holland and Naomi Quinn, eds. Cambridge: Cambridge University Press.

Lamaze, Fernand 1965 *Painless Childbirth: Psychoprophylactic Method.* New York: Pocket Books.

Laqueur, Thomas 1986 "Female Orgasm, Generation, and the Politics of Reproductive Biology." *Representations* 14(Spring):1–82.

Lasch, Christopher 1984 *The Minimal Self: Psychic Survival in Troubled Times.* New York: W. W. Norton.

Lauersen, Niels H. and Eileen Stukane 1983 *PMS Premenstrual Syndrome and You: Next Month Can Be Different.* New York: Simon and Schuster.

Lauritzen, Christian and Pieter A. van Keep 1978 *Estrogen Therapy: The Benefits and Risks. Frontiers of Hormone Research,* V. 5, Tj. B. van Wimersma Greidanus, ed. Basel: S. Karger.

Laws, Sophie 1983 "The Sexual Politics of Pre-Menstrual Tension." *Women's Studies International Forum* 6(1):19–31.

Lein, Allen 1979 *The Cycling Female: Her Menstrual Rhythm.* San Francisco: W. H. Freeman.

Lerner, Monroe and Richard N. Stutz 1975 "Mortality Differentials Among Socioeconomic Strata in Baltimore, 1960 and 1973," in *Proceedings of the American Statistical Association,* 517–22.

Lever, Judy with Dr. Michael G. Brush 1981 *Pre-menstrual Tension.* New York: Bantam.

Levine, Louis 1924 *The Women's Garment Workers: A History of the International Ladies' Garment Workers' Union.* New York: B. W. Huebsch, Inc.

Levit, Leonore 1963 *Anxiety and the Menopause: A Study of Normal Women.* Unpublished Ph.D. dissertation, University of Chicago.

Lewis, I. M. 1971 *Ecstatic Religion.* Middlesex, England: Penguin.

Lewontin, R. C., Steven Rose, and Leon J. Kamin 1984 *Not in Our Genes: Biology, Ideology, and Human Nature.* New York: Pantheon.

Lock, Margaret 1982 "Models and Practice in Medicine: Menopause as Syndrome or Life Transition?" *Culture, Medicine and Psychiatry* 6:261–80.

———. 1986 "Ambiguities of Aging: Japanese Experience and Perceptions of Menopause." *Culture, Medicine, and Psychiatry* 10(1):23–46.

Lomax, P. 1982 "The Pathophysiology of Postmenopausal Hot Flushes." *Reproduccion* 6:93–99.

Lorde, Audre 1981 "The Uses of Anger." *Women's Studies Quarterly* 9(3):7–10.

———. 1982 *Chosen Poems: Old and New.* New York: W. W. Norton.

Luker, Kristin 1984 *Abortion and the Politics of Motherhood.* Berkeley: University of California Press.

Mackintosh, Maureen 1977 "Reproduction and Patriarchy: A Critique of Claude Meillassoux, 'Femmes, Greniers et Capitaux.'" *Capital and Class* 2:119–127.

MacPherson, Kathleen 1981 "Menopause as Disease: The Social Construction of a Metaphor." *Advances in Nursing Science* 3(2):95–113.

Mann, Michael 1973 *Consciousness and Action among the Western Working Class.* London, England: The Macmillan Press, Ltd.

Mannheim, Karl 1936 *Ideology and Utopia: An Introduction to the Sociology of Knowledge.* New York: Harcourt Brace Jovanovich.

Manning, Peter K. and Horacio Fabrega, Jr. 1973 "The Experience of Self and Body: Health and Illness in the Chiapas Highlands," pp. 251–301 in *Phenomenological Sociology: Issues and Applications,* George Psathas, ed. New York: John Wiley.

Marcuse, Herbert 1964 *One-Dimensional Man: Studies in the Ideology of Advanced Industrial Society.* Boston: Beacon Press.

Marieskind, Helen I. 1979 *An Evaluation of Caesarean Section in the United States.* Washington, DC: Department of Health, Education and Welfare.

———. 1983 "Cesarean Section." *Women and Health* 7(3–4):179–98.

Marut, Joanne Sullivan 1978 "The Special Needs of the Cesarean Mother." *The American Journal of Maternal Child Nursing,* July/August:202–206.

Marut, Joanne Sullivan and Ramona T. Mercer 1979 "Comparison of Primiparas' Perceptions of Vaginal and Cesarean Births." *Nursing Research* 28(5):260–66.

Marx, Karl 1967a *Capital: A Critique of Political Economy.* V. 1. New York: International Publishers.

———. 1967b *Capital: A Critique of Political Economy,* V. 3. Frederick Engels, ed. New York: International Publishers.

Maryland Center for Health Statistics 1984 *Health Maryland 1984.* Baltimore: Maryland Department of Health and Mental Hygiene.

Mason, Elliott B. 1983 *Human Physiology.* Menlo Park, CA: Benjamin Cummings Publishing Co.

Matria, C. and Patricia Mullen 1978 "Reclaiming Menstruation: A Study of Alienation and Repossession." *Women and Health* 3(3):23–30.

McCance, R. A., M. C. Luff, and E. E. Widdowson 1937 "Physical and Emotional Periodicity in Women." *Journal of Hygiene* 37:571–614.

McCrea, Frances B. 1983 "The Politics of Menopause: The 'Discovery' of a Deficiency Disease." *Social Problems* 31(1):111–23.

McCrea, Frances B. and Gerald E. Markle 1984 "The Estrogen Replacement Controversy in the USA and UK: Different Answers to the Same Question?" *Social Studies of Science* 14:1–26.

McKinlay, Sonja M. and Margot Jefferys 1974 "The Menopausal Syndrome." *British Journal of Preventive Social Medicine* 28:108–15.

McNaught, Ann B. and Robin Callander 1983 *Illustrated Physiology.* 4th Edition. Edinburgh: Churchill Livingstone.

Mead, Margaret 1965 *And Keep Your Powder Dry.* New York: William Morrow Co.

Meigs, Charles 1879 *Females and Their Diseases.* Philadelphia: D. G. Brinton.

Meillassoux, Claude 1981 *Maidens, Meal and Money: Capitalism and the Domestic Community.* Cambridge: Cambridge University Press.

Melman, Seymour 1983 *Profits without Production.* New York: Knopf.

Merchant, Carolyn 1980 *The Death of Nature: Women, Ecology, and the Scientific Revolution*. San Francisco: Harper and Row.

Mesnard, Jacques 1753 *Le Guide des Accoucheurs*. Paris: De Bure Le Breton et Durand.

Meyer, Linda D. 1981 *The Cesarean Revolution: A Handbook for Parents and Professionals*. Edmonds, WA: C. Franklin Press.

Minkoff, Howard L. and Richard H. Schwarz 1980 "The Rising Cesarean Section Rate: Can It Safely Be Reversed?" *Obstetrics and Gynecology* 56(2):135–43.

Mitchell, Juliet 1971 *Woman's Estate*. New York: Vintage.

Montgomery, David 1979 *Workers' Control in America: Studies in the History of Work, Technology, and Labor Struggles*. Cambridge: Cambridge University Press.

Moore, Barrington 1973 *Reflections on the Causes of Human Misery and upon Certain Proposals to Eliminate Them*. Boston: Beacon Press.

Moran, Marilyn A., ed. 1983 *The New Nativity: A Newsletter for Do-It-Yourself Homebirth Couples,* no. 24. Leawood, Kansas: New Nativity Press.

———. 1984 *The New Nativity: A Newsletter for Do-It-Yourself Homebirth Couples,* no. 30. Leawood, Kansas: New Nativity Press.

Morrison, Toni 1973 *Sula*. New York: New American Library.

Mountcastle, Vernon B. 1980 *Medical Physiology*. 14th Edition, V. II. St. Louis, MO: C. V. Mosby Co.

Muhlenkamp, Ann F., Margaret M. Waller, and Ann E. Bourne 1983 "Attitudes Toward Women in Menopause: A Vignette Approach." *Nursing Research* 32 (1):20–23.

Mullings, Leith 1984 "Minority Women, Work, and Health," pp. 121–38 in *Double Exposure: Women's Health Hazards on the Job and at Home,* Wendy Chavkin, ed. New York: Monthly Review Press.

Mumford, Lewis 1967 *The Myth of the Machine: Technics and Human Development*. V. I. New York: Harcourt, Brace and World.

———. 1970 *The Myth of the Machine: The Pentagon of Power*. V. 2. New York: Harcourt, Brace and World.

Murphy, Julie 1984 "Egg Farming and Women's Future," pp. 68–75 in *Test-tube Women,* Rita Arditti, Renate Duelli Klein, and Shelley Minden, eds. London: Routledge and Kegan Paul, Ltd.

Nelson, Bryce 1983 "Bosses Face Less Risk Than the Bossed." *New York Times,* 3 April.

Netter, Frank H. 1965 *A Compilation of Paintings on the Normal and Pathologic Anatomy of the Reproductive System*. The CIBA Collection of Medical Illustrations, V. II. Summit, NJ: CIBA.

Neugarten, Bernice, L. and Ruth J. Kraines 1965 "'Menopausal Symptoms' in Women of Various Ages," 27(3):266–73.

Neugarten, Bernice L., Vivian Wood, Ruth J. Kraines, and Barbara Loomis 1968 "Women's Attitudes Toward the Menopause," pp. 195–200 in *Middle*

Age and Aging: A Reader in Social Psychology, Bernice L. Neugarten, ed. Chicago: University of Chicago Press.
Newton, Niles 1968 "The Effects of Disturbance on Labor." *American Journal of Obstetrics and Gynecology* 101:1096–102.
Niswander, Kenneth R. 1981 *Obstetrics: Essentials of Clinical Practice.* 2nd Edition. Boston: Little, Brown and Co.
Noble, David 1984 *The Forces of Production.* New York: Knopf.
Norris, Ronald V. 1984 *PMS: Premenstrual Syndrome.* New York: Berkeley Books.
Norwood, Christopher 1984 *How to Avoid a Cesarean Section.* New York: Simon and Schuster.
Novak, Emil 1941 "Gynecologic Problems of Adolescence." *Journal of the American Medical Association* 117:1950–53.
———. 1944 *Textbook of Gynecology.* 2nd Edition. Baltimore, MD: Williams and Wilkins Co.
Novak, Emil and Edmund Novak 1952 *Textbook of Gynecology.* Baltimore, MD: Williams and Wilkins Co.
Novak, Edmund, Georgeanna Seegar Jones, and Howard W. Jones 1965 *Novak's Textbook of Gynecology.* 7th Edition. Baltimore, MD: Williams and Wilkins Co.
O'Brien, Mary 1981 *The Politics of Reproduction.* Boston: Routledge and Kegan Paul, Ltd.
O'Driscoll, Kieran and Michael Foley 1983 "Correlation of Decrease in Perinatal Mortality and Increase in Cesarean Section Rates." *Journal of the American College of Obstetricians and Gynecologists* 61(1):1–5.
O'Driscoll, Kieran and D. Meagher 1980 *Active Management of Labour.* London: W. B. Saunders.
O'Neill, Daniel J. 1982 *Menopause and Its Effect on the Family.* Washington, DC: University Press of America.
Oakley, Ann 1974 *The Sociology of Housework.* New York: Pantheon.
———. 1979a *Becoming a Mother.* New York: Schocken.
———. 1979b "A Case of Maternity: Paradigms of Women as Maternity Cases." *Signs* 4(4):607–31.
———. 1984 *The Captured Womb: A History of the Medical Care of Pregnant Women.* Oxford: Basil Blackwell.
Odent, Michel 1981 "The Evolution of Obstetrics at Pithiviers." *Birth and Family Journal* 8(1):7–15.
———. 1984 *Birth Reborn.* New York: Pantheon.
Okely, Judith 1975 "Gypsy Women: Models in Conflict," pp. 55–86 in *Perceiving Women,* Shirley Ardener, ed. New York: John Wiley.
Ollman, Bertell 1976 *Alienation: Marx's Conception of Man in Capitalist Society.* Cambridge: Cambridge University Press.
Ong, Aihwa 1983 "Japanese Factories, Malay Workers: Industrialization and Sexual Metaphors in West Malaysia." Manuscript.
Ortner, Sherry 1974 "Is Female to Male as Nature Is to Culture?" pp. 67–87

in *Women, Culture and Society,* Michelle Zimbalist Rosaldo and Louise Lamphere, eds. Stanford: Stanford University Press.

Osherson, Samuel and Lorna AmaraSingham 1981 "The Machine Metaphor in Medicine," pp. 218–249 in *Social Contexts of Health, Illness and Patient Care,* Elliot G. Mishler, Lorna R. AmaraSingham, Stuart T. Hauser, Ramsay Liem, Samuel D. Osherson, and Nancy Waxler, eds. Cambridge: Cambridge University Press.

Ott, William J., Frank Ostapowics, and Jeanne Meurer 1977 "Analysis of Variables Affecting Perinatal Mortality: St. Louis City Hospital 1969–75." *Obstetrics and Gynecology* 49(4):481–85.

The Oxford English Dictionary 1933 V. 6. Oxford: Oxford University Press.

Paige, Karen E. 1971 "Effects of Oral Contraceptives on Affective Fluctuations Associated with the Menstrual Cycle." *Psychosomatic Medicine* 33(6):515–37.

Panuthos, Claudia 1984 *Transformation through Birth.* South Hadley, MA: Bergin and Garvey.

Parker, Cornelia Stratton 1922 *Working with the Working Woman.* New York: Harper and Bros.

Parlee, Mary 1973 "The Premenstrual Syndrome." *Psychological Bulletin* 80(6):454–65.

Peiss, Kathy 1983 "'Charity Girls' and City Pleasures: Historical Notes," pp. 74–87 in *Powers of Desire: The Politics of Sexuality,* Ann Snitow, Christine Stansell, and Sharon Thompson, eds. New York: Monthly Review Press.

Penny, Virginia 1870 *How Women Can Make Money.* Springfield, MA: D. E. Fisk.

Percival, Eleanor 1943 "Menstrual Disturbances as They May Affect Women in Industry." *The Canadian Nurse* 39:335–37.

Petchesky, Rosalind Pollack 1983 "Reproduction and Class Divisions among Women," pp. 221–42 in *Class, Race, and Sex: The Dynamics of Control,* Amy Swerdlow and Hanna Lessinger, eds. Boston: G. K. Hall & Co.

Peterson, Gayle 1984 *Birthing Normally: A Personal Growth Approach to Childbirth.* Berkeley, CA: Mindbody Press.

The PMS Connection 1982–84 *PMS Connection.* Madison, WI: PMS Action.

Porter, Enid 1969 *Cambridgeshire Customs and Folklore.* London: Routledge and Kegan Paul, Ltd.

Posner, Judith 1979 "It's All in Your Head: Feminist and Medical Models of Menopause (Strange Bedfellows)." *Sex Roles* 5:179–90.

Powers, Marla N. 1980 "Menstruation and Reproduction: An Oglala Case." *Signs* 6:54–65.

Price, Sally 1984 *Co-wives and Calabashes.* Ann Arbor: University of Michigan Press.

Pritchard, Jack A. and Paul C. MacDonald 1980 *Williams Obstetrics.* 16th Edition. New York: Appleton-Century Crofts.

Pritchard, Jack A., Paul C. MacDonald, and Norman F. Gant 1985 *Williams Obstetrics.* 17th Edition. Norwalk, CT: Appleton-Century-Crofts.

Ranney, Brooks 1973 "The Gentle Art of External Cephalic Rotation." *American Journal of Obstetrics and Gynecology* 116(2):239–51.

Rapp, Rayna 1982 "Family and Class in Contemporary America: Notes Toward an Understanding of Ideology," pp. 168–87 in *Rethinking the Family: Some Feminist Questions,* Barrie Thorne with Marilyn Yalom, eds. New York: Longman.

Reitz, Rosetta 1977 *Menopause: A Positive Approach.* New York: Penguin.

Rich, Adrienne 1976 *Of Woman Born.* New York: Bantam.

———. 1978 "Transcendental Etude," in *The Dream of a Common Language: Poems 1974–1977.* New York: W. W. Norton.

Richardson, Dorothy 1905 "The Long Day: The Story of a New York Working Girl," pp. 1–303 in *Women at Work.* William O'Neill, ed. (Reprint Edition 1972, Quadrangle Books, Chicago.)

Robinson, Kathleen, Kathleen M. Huntington, and M. G. Wallace 1977 "Treatment of the Premenstrual Syndrome." *British Journal of Obstetrics and Gynaecology* 84:784–88.

Rosenberg, Charles E. 1979 "The Therapeutic Revolution: Medicine, Meaning, and Social Change in Nineteenth-Century America," pp. 3–25 in *The Therapeutic Revolution: Essays in the Social History of American Medicine,* Morris J. Vogel and Charles E. Rosenberg, eds. Philadelphia: University of Pennsylvania Press.

Rosengren, William R. and DeVault, Spencer 1963 "The Sociology of Time and Space in an Obstetrical Hospital," pp. 266–92 in *The Hospital in Modern Society,* Eliot Freidson, ed. New York: Free Press.

Rossi, Alice S. and Peter E. Rossi 1977 "Body Time and Social Time: Mood Patterns by Menstrual Cycle Phase and Day of the Week." *Social Science Research* 6:273–308.

Rothman, Barbara Katz 1982 *In Labor: Women and Power in the Birthplace.* New York: W. W. Norton.

———. 1984 "The Meanings of Choice in Reproductive Technology," pp. 23–33 in *Test-tube Women,* Rita Arditti, Renate Duelli Klein, and Shelley Minden, eds. London: Routledge and Kegan Paul, Ltd.

Rothman, Robert A. 1978 *Inequality and Stratification in the United States.* Englewood Cliffs, NJ: Prentice-Hall.

Rothman, Sheila M. 1978 *Woman's Proper Place.* New York: Basic Books.

Rothschild, Joan 1981 "Technology, 'Women's Work' and the Social Control of Women," pp. 160–83 in *Women, Power and Political Systems,* Margherita Rendel with Georgina Ashworth, eds. New York: St. Martin's Press.

Rothstein, William G. 1972 *American Physicians in the Nineteenth Century: From Sects to Science.* Baltimore: Johns Hopkins University Press.

Rowbotham, Sheila 1973 *Woman's Consciousness, Man's World.* New York: Penguin Books.

Royall, Nicki 1983 *You Don't Have to Have a Repeat Cesarean.* New York: Frederick Fell.

Rubin, Lillian Breslow 1976 *Worlds of Pain: Life in the Working-Class Family.* New York: Basic Books.

Russell, J. K. 1982 "Breech: Vaginal Delivery or Caesarean Section?" *British Medical Journal* 285:830–31.

Ryle, Gilbert 1968 "The Thinking of Thoughts: What Is 'Le Penseur' Doing?" in *Collected Papers,* V. II, *Collected Essays 1929–1968.* London: Hutchinson & Co., Ltd.

Sanders, Lawrence 1981 *The Third Deadly Sin.* New York: Berkley Books.

Schilder, Paul 1935 *The Image and Appearance of the Human Body.* Psyche Monographs No. 4. London: Kegan Paul, Trench, Trubner.

Schneider, David M. 1969 "Kinship, Nationality and Religion in American Culture: Toward a Definition of Kinship," pp. 116–25 in *Forms of Symbolic Action: Proceedings of the 1969 Annual Spring Meeting of the American Ethnological Society,* Robert F. Spencer, ed. Seattle: University of Washington Press.

Schneider, David M. and Raymond T. Smith 1973 *Class Differences and Sex Roles in American Kinship and Family Structure.* Englewood Cliffs, NJ: Prentice-Hall.

Scott, W. Clifford M. 1949 "The 'Body Scheme' in Psychotherapy." *British Journal of Medical Psychology* 22:139–150.

Scott, Joan Wallach 1980 "The Mechanization of Women's Work." *Scientific American,* March: 167–85.

Scully, Diana 1980 *Men Who Control Women's Health: The Miseducation of Obstetrician-Gynecologists.* Boston: Houghton Mifflin.

Seeman, Melvin 1975 "Alienation Studies." *Annual Review of Sociology* 1:91–123.

Segal, Daniel 1984 "Playing Doctor, Seriously: Graduation Follies at an American Medical School." *International Journal of Health Services* 14(3):379–96.

Seifer, Nancy 1976 *Nobody Speaks for Me!: Self-Portraits of American Working Class Women.* New York: Simon and Schuster.

Sennett, Richard and Jonathan Cobb 1972 *The Hidden Injuries of Class.* New York: Vintage.

Sernka, Thomas and Eugene Jacobson 1983 *Gastrointestinal Physiology: The Essentials.* Baltimore: Williams and Wilkins.

Severne, L. 1979 "Psycho-Social Aspects of the Menopause," pp. 101–20 in *Psychosomatics in Peri-Menopause.* A. A. Haspels and H. Musaph, eds. Baltimore: University Park Press.

Seward, G. H. 1934 "The Female Sex Rhythm." *Psychological Bulletin* 31: 153–192.

———. 1944 "Psychological Effects of the Menstrual Cycle on Women Workers." *Psychological Bulletin,* V. 41:90–102.

Shapiro, Sam, Edward R. Schlesinger, and Robert E. L. Nesbitt 1968 *Infant, Perinatal, Maternal and Childhood Mortality in the United States.* Cambridge: Harvard University Press.

Shapiro-Perl, Nina 1984 "Resistance Strategies: The Routine Struggle for Bread and Roses," pp. 193–208 in *My Troubles Are Going to Have Trouble with Me: Everyday Trials and Triumphs of Women Workers.* Karen Brodkin Sacks and Dorothy Remy, eds. New Brunswick, NJ: Rutgers University Press.

Shepard, Jon M. 1971 *A Study of Office and Factory Workers.* Cambridge, MA: MIT Press.

———. 1977 "Technology, Alienation, and Job Satisfaction." *Annual Review of Sociology* 3:1–21.

Shorter, Edward 1982 *A History of Women's Bodies.* New York: Basic Books.

Shuttle, Penelope and Peter Redgrove 1978 *The Wise Wound: Eve's Curse and Everywoman.* New York: Richard Marek.

Skultans, Vieda 1970 "The Symbolic Significance of Menstruation and the Menopause." *Man* 5(4):639–51.

———. 1985 "Vicarious Menstruation." *Social Science and Medicine* 21(6): 713–14.

Smith, Dorothy E. 1979 "A Sociology for Women," pp. 135–87 in *The Prism of Sex: Essays in the Sociology of Knowledge,* Julia Sherman and Evelyn Torton Beck, eds. Madison: University of Wisconsin Press.

Smith-Rosenberg, Carroll 1974 "Puberty to Menopause: The Cycle of Femininity in Nineteenth-century America," pp. 23–37 in *Clio's Consciousness Raised,* Mary Hartman and Lois W. Banner, eds. New York: Harper and Row.

Snijders, P. 1973 "Seventy-two years of Caesarean Section in One Hospital," pp. 83–91 in *Perinatal Medicine,* H. Bossart, J. M. Crus, A. Huber, L. S. Prod'hom, J. Sistek, eds. Third European Congress of Perinatal Medicine, Lausanne, 1972. Vienna: Hans Huber.

Sokolov, Natalie 1984 "Women in the Professions," lecture in series "Women in History, Culture and Society," Baltimore.

Sommer, Barbara 1973 "The Effect of Menstruation on Cognitive and Perceptual-Motor Behavior: A Review." *Psychosomatic Medicine* 35(6):515–34.

Sontag, Susan 1979 *Illness as Metaphor.* New York: Vintage.

Southam, Anna L. and Florante P. Gonzaga 1965 "Systemic Changes During the Menstrual Cycle." *American Journal of Obstetrics & Gynecology* 91(1):142–65.

Special Task Force to the Secretary of Health, Education, and Welfare 1973 *Work in America.* Cambridge, MA: MIT Press.

Speert, Harold 1973 *Iconographia Gyniatrica: A Pictorial History of Gynecology and Obstetrics.* Philadelphia: F. A. Davis.

Stack, Carol 1974 *All Our Kin: Strategies for Survival in a Black Community.* New York: Harper and Row.

Starr, Paul 1982 *The Social Transformation of American Medicine.* New York: Basic Books.

Steinem, Gloria 1981 *Outrageous Acts and Everyday Rebellions.* New York: Holt, Rinehart and Winston.

Strathern, Marilyn 1984 "Domesticity and the Denigration of Women," pp. 13–31 in *Rethinking Women's Roles: Perspectives from the Pacific,* Denise O'Brien & Sharon W. Tiffany, eds. Los Angeles: University of California Press.

Sullerot, Evelyne 1971 *Women, Society, and Change.* New York: World University Library.

Sutton, Constance, Susan Makiesky, Daisy Dwyer, and Laura Klein 1975 "Women, Knowledge, and Power," pp. 581–600 in *Women Cross-Culturally: Change and Challenge,* Ruby Rohrlich-Leavitt, ed. The Hague: Mouton.

Taub, Nadine and Elizabeth M. Schneider 1982 "Perspectives on Women's Subordination and the Role of Law," pp. 117–39 in *The Politics of Law: A Progressive Critique,* David Kairys, ed. New York: Random House.

Taussig, Michael T. 1980 *The Devil and Commodity Fetishism in South America.* Chapel Hill: University of North Carolina Press.

Taylor, J. Madison 1904 "The Conservation of Energy in Those of Advancing Years." *Popular Science Monthly,* 64:343–414, 541–49.

Therborn, Göran 1980 *The Ideology of Power and the Power of Ideology.* London: Editions and NLB.

Thompson, E. P. 1967 "Time, Work-Discipline, and Industrial Capitalism." *Past and Present,* 38:56–97.

Tilly, Louise A. and Joan W. Scott 1978 *Women, Work and Family.* New York: Holt, Rinehart and Winston.

Tilt, Edward John 1857 *The Change of Life in Health and Disease.* London: John Churchill.

Todd, Linda 1981 *Labor and Birth: A Guide for You.* Minneapolis: International Childbirth Education Association.

Topley, Marjorie 1975 "Marriage Resistance in Rural Kwangtung," pp. 67–88 in *Women in Chinese Society,* Margery Wolf and Roxane Witke, eds. Stanford: Stanford University Press.

Trowell, Judith 1982 "Possible Effects of Emergency Caesarean Section on the Mother-Child Relationship." *Early Human Development* 7:41–51.

Ullery, J. C. and M. A. Castallo 1957 *Obstetric Mechanisms and Their Management.* Philadelphia: F. A. Davis.

U.S. Department of Health and Human Services 1981 *Cesarean Childbirth.* Washington, DC: NIH Publication No. 82–2067.

van Keep, Pieter A. and Jean M. Kellerhals 1974 "The Impact of Socio-Cultural Factors on Symptom Formation: Some Results of a Study on Ageing Women in Switzerland." *Psychotherapy and Psychosomatice* 23:251–63.

Van Vorst, Mrs. John and Marie 1903 *The Woman Who Toils: Being the Experience of Two Gentlewomen as Factory Girls.* New York: Doubleday, Page and Co.

Vander, Arthur J., James H. Sherman, and Dorothy S. Luciano 1980 *Human*

Physiology: The Mechanisms of Body Function. 3rd Edition. New York: McGraw-Hill.

———. 1985 *Human Physiology: The Mechanisms of Body Function*. 4th Edition, New York: McGraw-Hill.

Vanek, Joan 1974 "Time Spent in Housework." *Scientific American* 231(5):116–20.

Venable, Vernon 1945 *Human Nature: The Marxian View*. New York: Knopf.

Vidius, Vidus 1611 *Ars Medicinalis*. 3 Vols. Venice: Juntae.

Völter, Christoph 1687 *Neueröffnette Hebammen-Schul*. Stuttgart: Johann Gottfried Zubrodt.

Watkins, Linda M. 1986 "Premenstrual Distress Gains Notice as a Chronic Issue in the Workplace." *Wall Street Journal*, 22 Jan.

Walton, Vicki E. 1976 *Have It Your Way*. New York: Bantam.

Weekes, Leroy R. 1983 "Cesarean Section: A Seven-Year Study." *Journal of the National Medical Association* 75(5):465–76.

Wegman, Myron E. 1984 "Annual Summary of Vital Statistics, 1983." *Pediatrics* 74(6):981–90.

Weideger, Paula 1977 *Menstruation and Menopause: The Physiology and Psychology, the Myth and the Reality*. New York: Delta.

Weinberg, Martin S. 1968 "Embarrassment: Its Variable and Invariable Aspects." *Social Forces* 46:382–88.

Weindler, F. 1908 *Geschichte der Gynakologisch-Anatomischen Abbildung*. Dresden: Zahr and Jaensch.

Weiner, Lynn Y. 1985 *From Working Girl to Working Mother: The Female Labor Force in the United States, 1820–1980*. Chapel Hill: University of North Carolina Press.

Wertz, Richard W. and Dorothy L. Wertz 1977 *Lying-in: A History of Childbirth in America*. New York: Schocken.

Westfall, Richard S. 1971 *The Construction of Modern Science: Mechanisms and Mechanics*. Cambridge: Cambridge University Press.

Whisnant, Lynn, Elizabeth Brett, and Leonard Zergans 1975 "Implicit Messages Concerning Menstruation in Commercial Educational Materials Prepared for Young Girls." *American Journal of Psychiatry* 132(8):815–20.

Whistnant, Lynn and Leonard Zergans 1975 "A Study of Attitudes Toward Menarche in White Middle-Class American Adolescent Girls." *American Journal of Psychiatry* 132(8):809–14.

Whorf, Benjamin Lee 1956 *Language, Thought, and Reality: Selected Writings of Benjamin Lee Whorf*. Cambridge, MA: MIT Press.

Wilbush, Joel 1981 "What's in a Name? Some Linguistic Aspects of the Climacteric." *Maturitas* 3:1–9.

Williams, Raymond 1979 *Politics and Letters: Interviews with New Left Review*. London: New Left Review.

Wilson, Christine Coleman and Wendy Roe Hovey 1980 *Cesarean Childbirth: A Handbook for Parents*. New York: Doubleday.

Winner, Langdon 1977 *Autonomous Technology: Technics-out-of-Control as a Theme in Political Thought*. Cambridge: MIT Press.

Witkowski, G. J. 1891 *Accoucheurs et Sages-Femmes Célèbres*. Paris: G. Steinheil.

Witt, Reni L. 1984 *PMS: What Every Woman Should Know about Premenstrual Syndrome*. New York: Stein and Day.

Wittgenstein, Ludwig 1969 *On Certainty*, G. E. M. Anscombe & G. H. von Wright, eds. New York: Harper & Row.

Wood, Ann Douglas 1973 "'The Fashionable Diseases': Women's Complaints and Their Treatment in Nineteenth-Century America." *Journal of Interdisciplinary History* 4(1):25–52.

World Health Organization Scientific Group 1981 *Research on the Menopause*. World Health Organization Technical Report Series 670. Geneva: World Health Organization.

Young, Diony 1982 *Changing Childbirth: Family Birth in the Hospital*. Rochester, New York: Childbirth Graphics.

Zaretsky, Eli 1976 *Capitalism, the Family and Personal Life*. New York: Harper and Row.

Index

Disclaimer: The names of all persons quoted from interviews in this book are pseudonyms. In no case is an interviewee's real name used as the source of an interview statement. Any resemblance between the pseudonyms and the name of any person living or dead is entirely coincidental, and should not be construed as implying a connection between the interviewee and any other person.

EMILY MARTIN is Mary Garrett Professor of Anthropology at the Johns Hopkins University and author of *The Cult of the Dead in a Chinese Village* and *Chinese Ritual and Politics*.